Birkhäuser

Modern Birkhäuser Classics

Many of the original research and survey monographs in pure and applied mathematics published by Birkhäuser in recent decades have been groundbreaking and have come to be regarded as foundational to the subject. Through the MBC Series, a select number of these modern classics, entirely uncorrected, are being re-released in paperback (and as eBooks) to ensure that these treasures remain accessible to new generations of students, scholars, and researchers.

Klaus Bichteler

Integration – A Functional Approach

Reprint of the 1998 Edition

Klaus Bichteler
Department of Mathematics
The University of Texas
Austin, TX 78712
USA
kbi@math.utexas.edu

1991 Mathematics Subject Classification 28-01, 28C05

ISBN 978-3-0348-0054-9 e-ISBN 978-3-0348-0055-6
DOI 10.1007/978-3-0348-0055-6

Originally published under the same title in the Birkhäuser Advanced Texts – Basler Lehrbücher series
by Birkhäuser Verlag, Switzerland, ISBN 978-3-7643-5936-2
Reprinted 2010 by Springer Basel AG

Cover design: deblik, Berlin

Printed on acid-free paper

Springer Basel AG is part of Springer Science+Business Media

www.birkhauser-science.com

For Ursula

Contents

Preface

This text originated as notes for the introductory graduate Real Analysis course at the University of Texas at Austin. They were intended for students who have had a first course in Real Analysis (ϵ–δ–proofs) and know the topology of the real line, in particular the notion of compactness, and the Riemann integral.

The main topic in such a course traditionally is integration theory, and so it is here. The approach taken is different from the usual one, though. Usually the development follows history. First the content is extended to a measure on a σ–algebra, — which is identified through Carathéodory's cut condition — then to the integrable functions through another set of limit operations. Daniell observed that the extensions have much in common and can be done in one effort, saving half the labor. We follow in his footsteps. The emphasis is not, however, on saving time so much as it is on functionality: we ask what the purpose and expected properties of the prospective integral are and then fashion definitions and procedures accordingly—history and the usual treatment are reviewed, though, in Section III.7, so as to give the reader the connection of the historical development with the present one.

Functionality also guides the treatment of measurability. The text does not set down the notion of a σ–algebra as a gift from our betters, followed by a (mysterious) definition of measurability on them; this would pressure the gentle reader into accepting authoritarian mathematics or leave the rebellious one with many questions: where do these definitions come from, where do they lead, and how did anyone ever dream them up? Rather, we ask what the obstruction to the integrability of a function might be besides being too big, in other words, what the local structure of the integrable functions is. Two simple observations in answer to this lead to Littlewood's Principles, which are used to *define* measurability. The permanence properties of measurability are rather easy to establish this way, and once they are available so is the fact that the definition agrees with the traditional one.

A third uncommon aspect of the development is the use of seminorms. Instead of arguing with the upper and lower integrals of Daniell, analogs of Lebesgue's outer and inner measure, we observe that already in the case of the Riemann integral a simple absolute–value sign under the upper integral simplifies matters considerably, by rendering superfluous the study of the lower integral and providing a seminorm, the Daniell mean. The analyst will often attack a problem by defining a suitable seminorm with which to identify a function space in which a solution reasonably can be sought. It seems to me that it is not too early to introduce this tool in the context of integration: the problem is to extend the elementary integral from step functions to larger classes, and the Jordan and Daniell means are perfectly suitable seminorms towards that goal.

The exposition starts with a review of Weierstraß' theorem and the Riemann integral. The latter review is designed to show that the integral of Lebesgue–Carathéodory emerges from rather a minute alteration to Riemann's; it is not the mystery but the plethora of wonderful properties of the new integral that ask for another 120 pages just to write them down and prove them.

The exercises are an integral part of the material. Those marked by an asterisk* are used later on and thus must be done. For the majority of them there are answers or at least hints in the appendix.

The first version of these notes was replete with mistakes, and I thank my students for pointing them out. I hope the remaining mistakes are small in number; they are all mine.

Klaus Bichteler
The University of Texas at Austin
//www.ma.utexas.edu/users/kbi
kbi@math.utexas.edu
May 1998

Chapter I

Review

I.1 Introduction

Riemann's integral is a major achievement of civilization. It solves in a rigourous way the ancient problems of squaring the disk and straightening the circle. It provides a solid foundation for Newtonian Physics, on which modern theories of physics and engineering can grow. Nevertheless, it has shortcomings: its limit theorems are not good enough, and not sufficiently many functions are Riemann–integrable. Consequently it is an insufficient tool for quantum physics and the associated partial differential equations. These shortcomings can be overcome, though, by a careful analysis of its concepts and a surprisingly slight improvement on them.

The integral of a step function whose steps are intervals is the sum of the products *step–size* $\times$ *height–of–step*. Riemann's well–known procedure to integrate a function f that is not a step function is to squeeze it with step–functions from above and below: One defines the upper Riemann integral $\int^{\natural}$ of f to be the infimum of the integrals of step–functions above f and its lower Riemann integral $\int_{\natural}$ to be the supremum of the integrals of step–functions below. Then one says that f is Riemann integrable if these two numbers agree, and in that case defines the Riemann integral of f as the common value. We shall refer to this idea as the *"Riemann squeeze."*

Lebesgue's original idea towards enlarging the class of integrable functions was this: if at the outset more sets than just intervals could be measured then the class of step functions, to which an integral could be assigned by the formula sum of the products *step–size* $\times$ *height–of–step*, would be richer; more functions would be available to do the squeezing with; thus the Riemann–squeeze could be applied to more functions. This would result in a larger class of integrable functions.

Lebesgue did, indeed, succeed in measuring many more sets. It turns out, however, that the Riemann squeeze by itself is not adequate even then, and that further limit operations have to be performed in order to turn this idea into a complete success — the main reason being that the Riemann construction of the upper and lower integrals of f still requires the absolute value $|f|$ of the integrand f to be majorized by a step function, i.e., to be bounded and to vanish outside some integrable set.

K. Bichteler, *Integration – A Functional Approach*, Modern Birkhäuser Classics,
DOI 10.1007/978-3-0348-0055-6_1, © Birkhäuser Verlag 1998

Daniell noticed that Lebesgue's original extension of the measure from intervals to his larger class of integrable sets and the subsequent limit operations had much in common and that half the labour can be saved by combining the two efforts. Daniell's method consists in replacing the Riemann upper and lower integrals by slightly different ones, termed the Daniell upper and lower integrals $\int^*$ and $\int_*$, and then to squeeze as before. We shall follow his path to Lebesgue's integral, modified minutely by the following observation:

A Riemann integrable function f is evidently close to being a step function itself, since it can be squeezed arbitrarily closely by the latter. This somewhat vague statement can be given a precise meaning by the following notion of the natural distance of f from a step function ϕ:

$$\operatorname{dist}^\natural(f,\phi) = \|f-\phi\|^\natural = \int^\natural |f-\phi| \,.$$

A simple application of the triangle inequality shows that f is Riemann integrable if and only if there is a sequence (ϕ_n) of step functions whose distance $\|f-\phi_n\|^\natural$ from f goes to zero. The integral of f is then the limit of the integrals of the ϕ_n. The size measurement $\|\ \|^\natural$ is a *seminorm.* Similarly, Lebesgue integrable are precisely those functions f whose distance from step functions ϕ as measured by the *Daniell mean* $\|\ \|^*$,

$$\operatorname{dist}^*(f,\phi) = \|f-\phi\|^* = \int^* |f-\phi| \,,$$

can be made as small as one pleases. Again the integral is the extension by continuity. Looking at the integral this way, namely, as an extension by continuity under a suitable seminorm, reduces the technicalities some more and is in keeping with modern analysis, where seminorms and the notion of extension by continuity play an all–pervading rôle. Furthermore, it aims right at the heart of the problem: measure theory is, despite its name, not so much about measuring sets as about sizing functions. (This is not meant to disparage the measurement of sets; most frequently, a nuts–and–bolts problem about estimating the mean, or another gauge of the size, of a function f involves estimating the measure of the set of points where f is large or small or has some other property of interest — see for instance Sections V.4 and V.10.)

After some notation has been established, this chapter continues with a section on the theorem of Weierstraß. A short review of the Riemann integral follows. It is explained in detail how the Riemann integral can be viewed as the extension of a linear map under a suitable seminorm. We take the occasion to explain this important notion guided by that familiar example.

I.2 Notation

The Reals. The reader is of course familiar with the field $\mathbb{R}$ of real numbers. It is the one and only complete ordered field. It contains the natural numbers $\mathbb{N} = \{0, 1, 2, \ldots\}$, the ring of integers $\mathbb{Z} = \{\ldots, -2, -1, 0, 1, 2, \ldots\}$, and the field $\mathbb{Q}$ of rational numbers. The rationals are dense in the reals. $\mathbb{R}$ is also called the ***real line***. The reader not familiar with any of these notions should consult her textbook on undergraduate Real Analysis or [9, pp. 31–50]. Occasionally the complex number field $\mathbb{C}$ comes into the picture.

Positivity and Negativity. A real number a will be called ***positive*** if $a \geq 0$, and $\mathbb{R}_+$ denotes the set of positive reals. Similarly, a real–valued function f is positive if $f(x) \geq 0$ for all points x in its domain $\mathrm{dom}(f)$. If we want to stress that f be ***strictly positive***: $f(x) > 0 \quad \forall x \in \mathrm{dom}(f)$, we shall say so. It is clear what the words "***negative***" and "***strictly negative***" mean. If $\mathcal{F}$ is a collection of functions then $\mathcal{F}_+$ will denote the positive functions in $\mathcal{F}$, etc. Note that a positive function may be zero on a large set, in fact, everywhere. The statements "b ***exceeds*** a," "b ***is bigger than*** a," and "a ***is less than*** b" all mean "$a \leq b$;" modified by the word "strictly" they mean "$a < b$."

Exercise 2.1 To practice this convention let f, g be two functions having the same domain. Write the negation of the statement "g exceeds f."

The Larger or Smaller of Two Real Numbers a, b is customarily denoted by $a \vee b$ or $a \wedge b$, respectively. Similarly, the pointwise maximum or minimum of two functions f, g is denoted by $f \vee g$ and $f \wedge g$, respectively:

$$f \vee g\,(x) = \max\{f(x), g(x)\} \quad \text{or} \quad f \wedge g\,(x) = \min\{f(x), g(x)\}\ .$$

$|f|$ is the function $x \mapsto |f(x)|$; and $f_+ = f \vee 0$, $f_- = (-f) \vee 0$, so that $f = f_+ - f_-$. Suppose $\mathcal{F}$ is a family of functions. Then $\bigvee \mathcal{F} = \bigvee_{f \in \mathcal{F}} f$ is their pointwise supremum:

$$\Big(\bigvee \mathcal{F}\Big)(x) = \sup_{f \in \mathcal{F}} f(x)\ ;\ \text{and}\ \Big(\bigwedge \mathcal{F}\Big)(x) = \inf_{f \in \mathcal{F}} f(x)$$

is their pointwise infimum. Very frequently, the family $\mathcal{F}$ is countable: $\mathcal{F} = \{f_1, f_2, \ldots\}$. Then we write

$$\Big(\bigvee_{n \in \mathbb{N}} f_n\Big)(x) = \sup_{n \in \mathbb{N}} f_n(x) \text{ and } \Big(\bigwedge_n f_n\Big)(x) = \inf_n f_n(x)\ ,$$

respectively. Recall that the limit inferior and the limit superior of a sequence (a_n) of reals are defined as

$$\liminf_{n\to\infty} a_n = \lim_{N\to\infty} \inf_{n\ge N} a_n = \bigvee_{N\in\mathbb{N}} \bigwedge_{n\ge N} a_n \quad \text{and}$$

$$\limsup_{n\to\infty} a_n = \lim_{N\to\infty} \sup_{n\ge N} a_n = \bigwedge_{N\in\mathbb{N}} \bigvee_{n\ge N} a_n \ , \quad \text{respectively.}$$

Exercise 2.2 For all $a, b, c, r \in \mathbb{R}$, $(a\wedge b)+c = (a+c)\wedge(b+c)$ and $(a\vee b)+c = (a+c)\vee(b+c)$; $r(a\vee b) = ra\vee rb$ if $r \ge 0$ and $r(a\vee b) = ra\wedge rb$ if $r \le 0$; $a\wedge b + a\vee b = a+b$, $a_+ = \frac{1}{2}(|a|+a)$, and $a_- = \frac{1}{2}(|a| - a)$; $a \vee b = \frac{1}{2}(a + b + |a - b|)$ and $a \wedge b = \frac{1}{2}(a + b - |a - b|)$.

The Extended Real Line $\overline{\mathbb{R}}$ is the real line augmented by the two symbols $-\infty$ and $\infty = +\infty$:

$$\overline{\mathbb{R}} = \{-\infty\} \cup \mathbb{R} \cup \{+\infty\} \, .$$

We adopt the usual conventions concerning the arithmetic and order structure of the extended reals $\overline{\mathbb{R}}$:

$$-\infty < r < +\infty \quad \forall r \in \mathbb{R}\,; \quad |\pm\infty| = +\infty\,;$$
$$-\infty \wedge r = -\infty\,, \quad \infty \vee r = \infty \quad \forall r \in \mathbb{R}\,;$$
$$-\infty + r = -\infty\,, \quad +\infty + r = +\infty \quad \forall r \in \mathbb{R}\,;$$

$$r \cdot \pm\infty = \begin{cases} \pm\infty & \text{for } r > 0\,, \\ 0 & \text{for } r = 0\,, \\ \mp\infty & \text{for } r < 0\,; \end{cases} \qquad \pm\infty^p = \begin{cases} \pm\infty & \text{for } p > 0\,, \\ 1 & \text{for } p = 0\,, \\ 0 & \text{for } p < 0\,. \end{cases}$$

The symbols $\infty - \infty$ and $0/0$ are not defined; there is no way to do this without confounding the order or the previous conventions.
When a set S of reals is unbounded above we write $\sup S = \infty$, and $\inf S = -\infty$ when it is not bounded below. It is convenient to define the infimum of the empty set $\emptyset$ to be $+\infty$ and to write $\sup\emptyset = -\infty$. A function whose values may include $\pm\infty$ is often called a ***numerical function***.

Exercise 2.3* This exercise establishes a simple argument that is used over and over. The reader should be sure to understand it thoroughly. Let $a, b \in \overline{\mathbb{R}}$. Then $a \le b$ if and only if any one of the following two conditions obtains.
Whenever $s \in \mathbb{R}$ is a number strictly bigger than b then s is also bigger than a.
Whenever $s \in \mathbb{R}$ is a number bigger than b, then $s + \epsilon$ is bigger than a, for any $\epsilon > 0$.
Design a similar criterion in terms of numbers s smaller than a.

Exercise 2.4 This exercise allows the reader to check his vocabulary: injection, surjection, bijection, metric space, compactness, continuous maps, homeomorphism, inverse of a map. Set $\arctan \pm\infty = \pm\pi/2$ and define $d(x, y) = |\arctan x - \arctan y|$ for $x, y \in \overline{\mathbb{R}}$. Show that d is a metric under which $\overline{\mathbb{R}}$ is a complete and compact space homeomorphic with the closed unit interval $I = [0, 1]$. That is to say, there is a bijection $\Phi : \overline{\mathbb{R}} \to I$

such that both Φ and and its inverse Φ^{-1} are continuous.
Show that the real line is homeomorphic with the open unit interval $(0,1)$.

Some Set Notation. Thc ***intersection*** of two sets A, B is the set of points common to both and is denoted by $A \cap B$:

$$A \cap B = \{x : x \in A \text{ and } x \in B\}; \quad \text{and} \quad A \cup B = \{x : x \in A \text{ or } x \in B\}$$

is the ***union*** of A and B, the set of points belonging to either set. The ***relative complement*** of B in A, or ***set difference***, $A \backslash B$ is the set of points in A that do not belong to B:

$$A \backslash B = \{x \in A : x \notin B\}.$$

It is generally understood that all sets in question are subsets of a fixed set, the ***ambient set*** S. In this case $S \backslash B$ is called the ***complement*** of B and is denoted B^c. Thus $A \backslash B = A \cap B^c$ if both A and B are understood to be subsets of S. Suppose $\mathcal{A}$ is a family of sets. Then $\bigcup \mathcal{A} = \bigcup_{A \in \mathcal{A}} A$ is their union and $\bigcap \mathcal{A} = \bigcap_{A \in \mathcal{A}} A$ is their intersection. Very frequently, the family $\mathcal{A}$ is countable: $\mathcal{A} = \{A_1, A_2, \ldots\}$. Then we write

$$\bigcup \mathcal{A} = \bigcup_{n \in \mathbb{N}} A_n \text{ and } \bigcap \mathcal{A} = \bigcap_{n} A_n ,$$

respectively.

The Expressions $[f > r]$, $[f = r]$, $[f < r]$,$[f \in S]$ etc. are shorthand for the sets $\{x \in \text{dom}(f) : f(x) > r\}$, $\{x \in \text{dom}(f) : f(x) = r\}$, $\{x \in \text{dom}(f) : f(x) < r\}$, $\{x \in \text{dom}(f) : f(x) \in S\}$ etc., respectively. For instance,

$$[\liminf f_n > 0]$$

is the set of points x where $f_n(x) > \epsilon$ for some $\epsilon > 0$, eventually.

Exercise 2.5 Let (f_n) be a sequence of real–valued functions. The set $[f_n \not\to]$ of points at which (f_n) does not converge equals

$$\bigcup_{r \in \mathbb{N}} \bigcap_{p \in \mathbb{N}} \bigcup_{m,n > p} [|f_m - f_n| > 1/r] .$$

The Indicator Function of a subset A of some set S is denoted by 1_A and is defined by

$$1_A(x) = \begin{cases} 1 & \text{if } x \in A; \\ 0 & \text{if } x \in A^c = S \backslash A. \end{cases}$$

Some authors denote this function by I_A, others by χ_A or 1_A. Let us overcome this trifling inconsistency by a convention that will simplify handsomely the typography, and life in general:

Convention 2.6 ***A set and its indicator function have the same name.***

If A is a set then A is also the function that has the value 1 on A and 0 off A; also, $3A$ is the function with $3A(x) = 3$ if $x \in A$, $3A(x) = 0$ otherwise. For another example, $[\phi \le r]$ is not only the set $\{x : \phi(x) \le r\}$ but also the function that equals 1 at any point $x \in \mathrm{dom}(\phi)$ at which $\phi(x) \le r$, and $[\phi \le r](x) = 0$ elsewhere. It will always be plain from the context whether a symbol refers to a set or to its indicator function (or both). Still, since this convention is somewhat unconventional, a little practice is indicated:

Exercise 2.7 (i) Let (f_n) be a sequence of functions. Then the sequence $([f_n \to] \cdot f_n)$ converges everywhere. (ii) A function f is (the indicator function of) a set iff it is ***idempotent***: $f^2 = f$. In this case $f = [f = 1] = [f \ne 0]$. (iii) $\int (0,1](x) \cdot x^2\, dx = 1/3$.

Exercise 2.8 Let A_n be a collection of sets. Then $A_1 \setminus A_2 = A_1 - A_1 \wedge A_2$; the symmetric difference $A_1 \bigtriangleup A_2 = A_1 \setminus A_2 \cup A_2 \setminus A_1$ equals $|A_1 - A_2|$; $\bigcup_n A_n = \bigvee_n A_n$; and $\bigcap A_n = \bigwedge A_n$.

Exercise 2.9 Let f be a positive bounded function. Show that it is the uniform limit of the functions (f_n) with finitely many values defined by

$$f_n = \sum_{k=-2^{2n}}^{2^{2n}} k2^{-n} \cdot [k2^{-n} \le f < (k+1)2^{-n}] .$$

Vector Spaces, Algebras, and Vector Lattices of functions will occupy us a great deal during these notes. It is well to define these notions now:

Definition 2.10 *Let $\mathcal{E}$ be a collection of functions, all defined on the same set S. Then $\mathcal{E}$ is a* ***linear space****, or* ***vector space****, of functions if it is closed under taking pointwise linear combinations; that is, if*

$$\phi, \phi' \in \mathcal{E},\ r, r' \in \mathbb{R} \Longrightarrow r \cdot \phi + r' \cdot \phi' \in \mathcal{E} .$$

$\mathcal{E}$ is a ***ring*** *of functions if it is a vector space and is closed under taking pointwise products:*

$$\phi, \phi' \in \mathcal{E} \Longrightarrow \phi \cdot \phi' \in \mathcal{E}.$$

If $\mathcal{E}$ is a ring and contains the constant function 1 then it is called an ***algebra****. $\mathcal{E}$ is a* ***vector lattice*** *or* ***Riesz space*** *of functions if it is a vector space and is closed under taking pointwise maxima and minima:*

$$\phi, \phi' \in \mathcal{E} \Longrightarrow \phi \vee \phi' \in \mathcal{E} \ \text{ and } \ \phi \wedge \phi' \in \mathcal{E}.$$

*$\mathcal{E}$ is **closed under chopping** if it contains, with every function ϕ, also the **chopped function** $\phi \wedge 1$.*
*If $\mathcal{E}$ is both a ring and a vector lattice closed under chopping then it is called a **lattice ring**; a lattice ring containing the constants is a **lattice algebra**.*

Example 2.11 Let us justify this definition by exhibiting a plethora of function spaces that meet one or more of the specifications above:

The polynomials on $\mathbb{R}$ or on $\mathbb{R}^n$ form algebras, those with zero constant coefficient form rings. The set $C[\mathbb{R}^n]$ of continuous functions on $\mathbb{R}^n$ is a lattice algebra closed under chopping.

The set $C^k[\mathbb{R}^n]$ of k–times continuously differentiable functions on $\mathbb{R}^n$ is an algebra again, in fact, is a *subalgebra* of $C[\mathbb{R}^n]$. So is the collection $C^\infty[\mathbb{R}^n]$ of infinitely differentiable functions on $\mathbb{R}^n$.

The bounded continuous functions $C_b[\mathbb{R}^n]$ form a lattice algebra. The collection $C_b^k[\mathbb{R}^n]$ of functions with k bounded continuous derivatives forms an algebra but is not a lattice.

The collection $C_0[\mathbb{R}^n]$ of continuous functions that vanish at infinity is a lattice ring. So is the collection $C_{00}[\mathbb{R}^n]$ of continuous functions with compact support [1]. The collection $C_{00}^k[\mathbb{R}^n]$ of k–times continuously differentiable functions of compact support form a ring.

The step functions (Definition 4.2) form a lattice ring. So do the Riemann integrable functions (Theorem 6.3) and the bounded Lebesgue integrable functions (Theorem II.5.5). ∎

If the constant function 1 belongs to the vector lattice $\mathcal{E}$ then $\phi \wedge 1$ belongs to $\mathcal{E}$ with every $\phi \in \mathcal{E}$, and $\mathcal{E}$ is closed under chopping. It simplifies matters, of course, to have the constants in $\mathcal{E}$; yet this is frequently used only to make sure that the functions in $\mathcal{E}$ can be chopped. In the context of integration many different vector lattices of functions occur naturally. Most are closed under chopping without however containing the constants.

The reader has discovered nothing new in the last page; she knew all along that the pointwise infimum or supremum of two continuous functions is continuous or that the sum and product of two bounded differentiable functions are bounded and differentiable, etc. The point lay not in these facts, but in a slight change of perspective — with Definition 2.10 on p. 6 we decided to view the forest rather than the single trees: we note that the *totality of continuous functions* has the property of being closed under taking pointwise suprema and infima, etc.; arguments on that totality that use only this feature apply to any other totality of functions with the same property and are therefore available without further travail. This is the point of the Definitions 2.10.

---

[1] The ***carrier*** of a function ϕ is the set $[\phi \neq 0]$ of points where it does not vanish. The ***support*** $\operatorname{supp} f$ of a function f is the closure of its carrier.

Supplements and Additional Exercises

Exercise 2.12* If $\mathcal{E}$ is a ring, ϕ an element of $\mathcal{E}$, and P is a polynomial with zero constant term, then the composition $P \circ \phi$ belongs to $\mathcal{E}$. If $\mathcal{E}$ is an algebra then the conclusion persists even if the polynomial has non–zero leading term.

Exercise 2.13* Let $\mathcal{L}$ be a vector space of functions on some set. Show that $\mathcal{L}$ is a vector lattice if and only if it has any one or all of the following properties:
(i) it is closed under taking finite minima; (ii) it is closed under taking finite maxima; (iii) it contains with every function f its absolute value $|f|$; (iv) it contains with every function f its positive part f_+.

Exercise 2.14* (i) Suppose $\mathcal{E}$ is a vector space of functions closed under chopping. Show that for any $\phi \in \mathcal{E}$ and $0 < r \in \mathbb{R}$, $\phi \wedge r$ and $\phi \vee (-r)$ belong to $\mathcal{E}$ as well. (ii) Give examples of vector lattices containing 1. (iii) Give examples of vector lattices that are closed under chopping yet do not contain the constant functions.

Exercise 2.15 Given a set $\mathcal{E}$ of functions set $\mathcal{E} + \mathbb{R} = \{\phi + r : \phi \in \mathcal{E}, r \in \mathbb{R}\}$. Show: if $\mathcal{E}$ is a vector space (ring, vector lattice *closed under chopping*, lattice ring) then $\mathcal{E} + \mathbb{R}$ is a vector space (algebra, vector lattice, lattice algebra). If $\mathcal{E}$ contains the constants then $\mathcal{E} + \mathbb{R} = \mathcal{E}$.

Definition 2.16 *Let $\mathcal{E}$ be a vector space of functions on some set S. A subset $A \subset S$ is called* ***confined by*** $\mathcal{E}$ *or* $\mathcal{E}$***–confined*** *if there is a function $\phi \in \mathcal{E}$ so that $A \le \phi$, i.e., $\phi \ge 0$ exceeds 1 on A. A function f on S is $\mathcal{E}$–confined if its carrier $[f \ne 0]$ is $\mathcal{E}$–confined. We denote by $\mathcal{E}_{00}$ the $\mathcal{E}$–confined functions in $\mathcal{E}$, and say $\mathcal{E}$ is* ***self–confining*** *or* ***self–confined*** *if $\mathcal{E}_{00} = \mathcal{E}$.*

Exercise 2.17* The spaces $C_b[\mathbb{R}^n]$, $C_{00}[\mathbb{R}^n]$, and the spaces $\mathcal{E}[\mathbb{R}^n]$ and $\mathcal{E}[\mathbb{R}]$ of step functions are self–confining; so are all other examples in 2.11 that contain the constants. $C_0[\mathbb{R}^n]$ is not self–confining.

Exercise 2.18* If $\mathcal{E}$ is a vector lattice closed under chopping then $\mathcal{E}_{00}$ is dense in $\mathcal{E}$.

Exercise 2.19* Let $(S, \mathcal{E})$ and $(S', \mathcal{E}')$ two sets equipped with vector spaces of bounded functions. On the cartesian product $S \times S'$ consider the collection $\mathcal{E} \otimes \mathcal{E}'$ of finite sums

$$\sum_i \phi_i \otimes \phi_i' : (s, s') \mapsto \sum_i \phi_i(s)\phi_i'(s') .$$

If both $\mathcal{E}$ and $\mathcal{E}'$ are vector spaces (rings, algebras, self–confined) then so is their ***tensor product*** $\mathcal{E} \otimes \mathcal{E}'$. There are vector lattices whose tensor product is not a vector lattice.

I.3 The Theorem of Stone–Weierstraß

This section develops some results about the uniform approximation of functions. The main result, Weierstraß' theorem on the line, may have been covered in the student's ϵ–δ–proof class. Nevertheless, it might be well to review this section, as a few more facts are extracted from the usual arguments than is customary. The next fact, simple as its proof is, will be used repeatedly.

Lemma 3.1 (Dini's theorem) *Let S be a compact space — the reader not familiar with general topology may view S as a closed and bounded subset of the line or of $\mathbb{R}^n$ — and let (ϕ_n) be a sequence of continuous functions on S that increases pointwise to another continuous function ϕ. Then (ϕ_n) converges to ϕ uniformly on S.*

Proof. Let $\epsilon > 0$ be given. The sets $U_n = [\phi - \phi_n < \epsilon]$ form an increasing open cover of S, so one of them, U_N say, will contain S, since S is compact. Clearly whenever $n \geq N$ then uniformly for all $x \in S \subset U_n$

$$|\phi(x) - \phi_n(x)| \leq \phi(x) - \phi_N(x) < \epsilon .$$

∎

Exercise 3.2 Let S be a compact space and let Φ be an ***increasingly directed*** collection of continuous functions on S; that is to say, for any two functions $\phi, \phi' \in \Phi$ there is a function $\psi \in \Phi$ bigger than both ϕ and ϕ'. Assume that the pointwise supremum of Φ is also a continuous function $\overline{\phi}$. Show that Φ converges uniformly to ϕ in the sense that for every $\epsilon > 0$ there is a $\phi_\epsilon \in \Phi$ such that $\|\overline{\phi} - \phi\|_u < \epsilon$ for all $\phi \in \Phi$ with $\phi \geq \phi_\epsilon$.

Exercise 3.3 Give a list of theorems involving uniform convergence of sequences of functions and recall their proofs.

Two Simple Results on the Real Line of a deceptively special aspect lead the way to Weierstraß' theorem.

Lemma 3.4 *Let t be a real number. Let the sequence $(p_n(t))$ of numbers be defined inductively by*

$$p_0(t) = 0, \quad \textit{and} \quad p_{n+1}(t) = \frac{1}{2}\Big(t^2 + 2p_n(t) - \big(p_n(t)\big)^2\Big).$$

Then $p_n(t)$ is a polynomial in t^2 with zero constant term. Furthermore, for any $t \in [-1, 1]$ and $n \in \mathbb{N}$ we have $0 \leq p_n(t) \leq p_{n+1}(t)$, and $(p_n(t))$ converges uniformly for $t \in [-1, 1]$ to the absolute value $|t|$ of t.

Proof. $p_1(t) = \frac{1}{2}t^2$ is a polynomial in t^2, and if p_n is then clearly so is p_{n+1}. If p_n has zero constant term then so does p_{n+1}. Thus all the p_n are polynomials in t^2 and vanish at zero. Two easy manipulations result in

$$2\big(|t| - p_{n+1}(t)\big) = \big(2 - |t|\big)|t| - \big(2 - p_n(t)\big)p_n(t)$$

and

$$2\big(p_{n+1}(t) - p_n(t)\big) = t^2 - \big(p_n(t)\big)^2 .$$

Now $x \mapsto (2 - x)x = 2x - x^2$ is increasing on $[0, 1]$. If, by induction hypothesis, $0 \leq p_n(t) \leq |t|$ for $|t| \leq 1$ then $p_{n+1}(t)$ will satisfy the same inequality; as it is true for p_0, it holds for all the p_n. The second equation shows that $p_n(t)$ increases with n for $t \in [-1, 1]$. As this sequence is also bounded, it has a limit $p(t) \geq 0$.

$p(t)$ must satisfy $0 = t^2 - (p(t))^2$ and thus equals $|t|$. Due to Dini's theorem, the convergence is uniform on $[-1, 1]$. ∎

Corollary 3.5 *Let $[-M, M]$ be a symmetric closed bounded interval, $M > 0$. There exists a sequence $(P_n(t))$ of polynomials in t with zero constant term that are positive on $[-M, M]$ and converge increasingly and uniformly on this interval to the absolute value function $t \mapsto |t|$.*

There is also a sequence $\big(Q_n(t)\big)$ of polynomials with zero constant term that converges uniformly on $[-M, M]$ to the function $t \mapsto t \wedge 1$.

Proof. For the first claim simply set

$$P_n(t) = M \cdot p_n(t/M) .$$

Now to the second. As

$$t \wedge 1 = \frac{1}{2}\big(t + 1 - |t - 1|\big) ,$$

the choice

$$Q_n(t) = \frac{1}{2}\big(t + 1 - P_n(t - 1)\big) \tag{3.1}$$

leaps to mind. These polynomials in t certainly decrease to $t \wedge 1$ uniformly on $[-M, M]$, as $n \to \infty$. However, they do not vanish at $t = 0$ and have to be adjusted by subtracting the constant $Q_n(0) = \frac{1}{2}\big(1 - P_n(1)\big)$, which tends to zero as n increases: the description is met by the adjusted sequence

$$Q_n(t) = \frac{1}{2}\big(t + P_n(1) - P_n(t - 1)\big) .$$

∎

Lemma 3.6 *Let $[-M, M]$ be a bounded closed symmetric interval and let $0 < \epsilon < 1$. The supremum of the linear functions $t \mapsto 2k\epsilon t - k^2\epsilon^2$,*

$$\begin{aligned}\phi_\epsilon(t) := &\bigvee\{2k\epsilon t - k^2\epsilon^2;\ k \in \mathbb{Z} : |k| < M/\epsilon\} \\ = &\bigvee\{2k\epsilon t - (2k\epsilon t \wedge k^2\epsilon^2);\ k \in \mathbb{Z} : |k| < M/\epsilon\} ,\end{aligned}$$

is everywhere on $[-M, M]$ smaller than t^2 and bigger than $t^2 - \epsilon$.

Proof. Let $\phi_k(t) = 2k\epsilon t - (k\epsilon)^2$. This linear function satisfies

$$t^2 - \phi_k(t) = (t - k\epsilon)^2 \geq 0 \quad \forall t \text{ , and } t^2 - \phi_k \leq \epsilon^2 < \epsilon \text{ for } |t - k\epsilon| \leq \epsilon.$$

The supremum ϕ_ϵ of the ϕ_k is thus smaller than t^2 and bigger than $t^2 - \epsilon$, inasmuch as the intervals $\{t : |t - k\epsilon| < \epsilon\} = ((k-1)\epsilon, k\epsilon)$, $|k| \leq M/\epsilon$, cover $[-M, M]$. If $k = 0$ then $\phi_k = 0$ and so $\phi_\epsilon = \bigvee_k(\phi_k \vee 0)$. Now

$$\phi_k \vee 0 = (2k\epsilon t - k^2\epsilon^2) \vee 0 = 2k\epsilon t - (2k\epsilon t \wedge k^2\epsilon^2) ,$$

which proves the equality of the two expressions for ϕ_ϵ. ∎

The two lemmas above say that the functions $t \mapsto |t|$ and $t \mapsto t^2$ can be approximated uniformly on arbitralily large intervals $[-M, M]$ by polynomials in t or by finite suprema of linear functions in t, respectively. At first sight they look rather uninteresting, since the functions to be approximated are so simple. Yet they can be parlayed into powerful results about the uniform approximation of functions on arbitrary compact spaces. Before doing this let us introduce a quantitative gauge for the uniform closeness of functions:

The Uniform Norm attaches a numerical size to a function and measures precisely how far apart two functions are in the uniform sense. It is the first of many seminorms we shall meet.

Definition 3.7 *Let f be a real–valued function. The* ***uniform norm*** *of f is the number*

$$\|f\|_u = \sup\{|f(x)| : x \in \text{dom}(f)\}.$$

The ***uniform distance*** *of two functions f, g defined on the same set is the uniform norm $\|f - g\|_u$ of their difference.*

Let $\mathcal{E}$ be a collection of functions with the same domain S. A function f on S is a ***uniform limit*** *of functions in $\mathcal{E}$ if there is a sequence (ϕ_n) in $\mathcal{E}$ whose uniform distance from f tends to zero: $\|f - \phi_n\|_u \xrightarrow[n\to\infty]{} 0$.*
The collection of all uniform limits of $\mathcal{E}$ is the ***uniform closure*** *of $\mathcal{E}$ and is denoted by $\overline{\mathcal{E}}^u$ or simply $\overline{\mathcal{E}}$.*

Exercise 3.8* Let $\mathcal{F}$ be a vector space of bounded real–valued functions on some set. A sequence (f_n) of functions in $\mathcal{F}$ converges to $f \in \mathcal{F}$ uniformly if and only if $\|f - f_n\|_u \xrightarrow[n\to\infty]{} 0$. Show further that $\| \ \|_u$ is
subadditive : $\|f + f'\|_u \le \|f\|_u + \|f'\|_u$,
absolute–homogeneous : $\|r{\cdot}f\|_u = |r| \cdot \|f\|_u$,
submultiplicative : $\|f \cdot g\|_u \le \|f\|_u \cdot \|g\|_u$,
and ***solid*** : $|f| \le |g| \implies \|f\|_u \le \|g\|_u$.

Exercise 3.9* $\overline{\overline{\mathcal{E}}} \stackrel{\text{def}}{=} \overline{\overline{\mathcal{E}}^u}^u = \overline{\mathcal{E}}^u$.

Proposition 3.10 *Let $\mathcal{E}$ be a collection of bounded functions on some set.*
If $\mathcal{E}$ is a ring **or** *a vector lattice closed under chopping then $\overline{\mathcal{E}}$ is* **both** *a ring* **and** *a vector lattice closed under chopping. In other words, $\overline{\mathcal{E}}$ is then a lattice ring.*

Proof. Let $\overline{\phi}, \overline{\psi} \in \overline{\mathcal{E}}$. Then there are sequences $(\phi_n), (\psi_n)$ in $\mathcal{E}$ converging uniformly to $\overline{\phi}, \overline{\psi}$, respectively. By Exercise 3.8

$$\|(r\overline{\phi} + s\overline{\psi}) - (r\phi_n + s\psi_n)\|_u \le |r|\|\overline{\phi} - \phi_n\|_u + |s|\|\overline{\psi} - \psi_n\|_u \xrightarrow[n\to\infty]{} 0 ,$$

$$\| \, |\overline{\phi}| - |\phi_n| \, \|_u \le \|\overline{\phi} - \phi_n\|_u \xrightarrow[n\to\infty]{} 0 ,$$
$$\|\overline{\phi} \wedge 1 - \phi_n \wedge 1\|_u \le \|\overline{\phi} - \phi_n\|_u \xrightarrow[n\to\infty]{} 0 ,$$

and

$$\|\phi \cdot \psi - \phi_n \cdot \psi_n\|_u \leq \|(\overline{\phi} - \phi_n) \cdot \overline{\psi}\|_u + \|\phi_n \cdot (\overline{\psi} - \psi_n)\|_u$$
$$\leq \|\overline{\phi} - \phi_n\|_u \cdot \|\overline{\psi}\|_u + (\sup_n \|\phi_n\|_u) \cdot \|\overline{\psi} - \psi_n\|_u \xrightarrow[n\to\infty]{} 0 .$$

The first line exhibits a linear combinations of functions in $\overline{\mathcal{E}}$ as an element of $\overline{\mathcal{E}}$ again, since $r\phi_n + s\psi_n \in \mathcal{E}$. Assume that $\mathcal{E}$ is a vector lattice closed under chopping. The second and third lines show that then so is $\overline{\mathcal{E}}$ — see Exercise 2.13 on p. 8. The last line shows that if $\mathcal{E}$ is a ring then so is $\overline{\mathcal{E}}$.
Next let us assume that $\mathcal{E}$ is a ring, pick the ϕ_n so that $\|\overline{\phi} - \phi_n\|_u \leq 1/n$, and set $M = \sup_n \|\phi_n\|_u$. By Corollary 3.5 there are polynomials P_n such that $\big|\,|t| - P_n(t)\,\big| \leq 1/n$ for all $t \in [-M, M]$. Then for all x in the ambient space

$$\big||\phi(x)| - P_n(\phi_n(x))\big| \leq \big||\overline{\phi}(x)| - |\phi_n(x)|\big| + \big||\phi_n(x)| - P_n(\phi_n(x))\big| ,$$

and

$$\big\|\,|\overline{\phi}| - P_n \circ \phi_n\,\big\|_u \leq \big\|\,|\overline{\phi}| - |\phi_n|\,\big\|_u + \big\|\,|\phi_n| - P_n \circ \phi_n\,\big\|_u \leq 2/n \xrightarrow[n\to\infty]{} 0 .$$

Since the P_n can be chosen to have zero constant term, the functions $P_n \circ \phi_n$ belong to $\mathcal{E}$ (Exercise 2.12), showing that $|\overline{\phi}| \in \overline{\mathcal{E}}$. By Exercise 2.13 $\overline{\mathcal{E}}$ is a vector lattice. We leave it to the reader to devise a similar argument based on the second part of Corollary 3.5 and proving that $\overline{\mathcal{E}}$ is closed under chopping.
Finally, let us assume that $\mathcal{E}$ is a vector lattice closed under chopping. Let $\overline{\phi} \in \overline{\mathcal{E}}$ and $0 < \epsilon < 1$ be given. Set $M = \|\overline{\phi}\|_u + 1$. For every $x \in \operatorname{dom}\overline{\phi}$, the number

$$\overline{\phi}_\epsilon(x) = \bigvee_{|k|<M/\epsilon} \left(2k\epsilon\overline{\phi}(x) - (2k\epsilon\overline{\phi}(x) \wedge k^2\epsilon^2)\right)$$

of Lemma 3.6 is then as close as ϵ to $\overline{\phi}(x)^2$, i.e., $\|\overline{\phi}_\epsilon - \overline{\phi}^2\|_u < \epsilon$. As $\overline{\phi}_\epsilon$ belongs to $\overline{\mathcal{E}}$, it is itself arbitrarily close to a function in $\mathcal{E}$, in the uniform distance. Thus $\overline{\phi}^2 \in \overline{\mathcal{E}}$. ***Polarization***

$$\overline{\phi} \cdot \overline{\psi} = 1/2\left((\overline{\phi} + \overline{\psi})^2 - \overline{\phi}^2 - \overline{\psi}^2\right)$$

shows that $\overline{\mathcal{E}}$ is closed under taking products and thus is a ring. ∎

Theorem 3.11 (The Stone–Weierstraß Theorem) *Let S be a compact Hausdorff space—the reader not familiar with general topology may view S as a closed and bounded subset of $\mathbb{R}$ or of $\mathbb{R}^n$ — and $\mathcal{E}$ a ring,* **or** *a vector lattice closed under chopping, of continuous functions on S.*
Assume that $\mathcal{E}$ ***separates the points****; that is to say, for any two points $s \neq t$ in S there is a function $\phi \in \mathcal{E}$ with $\phi(s) \neq \phi(t)$.*

(i) If there is no point at which all the functions of $\mathcal{E}$ vanish then every continuous function f on S can be approximated uniformly on S by functions from $\mathcal{E}$.
(ii) If all the functions of $\mathcal{E}$ vanish at the point $z \in S$ then every continuous function f on S that vanishes at z can be approximated uniformly on S by functions in $\mathcal{E}$ —since $\mathcal{E}$ separates the points there is at most one such point z.

Proof. This means of course that, given $\epsilon > 0$, there is a function $\phi \in \mathcal{E}$ with

$$\|f - \phi\|_u = \sup\{|f(t) - \phi(t)| : t \in S\} < \epsilon .$$

Let $\mathcal{C}$ denote the collection of all continuous functions on S in case (i), of all continuous functions vanishing at z in case (ii). It is to be shown that $\overline{\mathcal{E}} = \mathcal{C}$. By Exercise 3.9 it suffices to show that $\mathcal{C}$ is the uniform closure of $\overline{\mathcal{E}}$. In other words, we may assume that $\mathcal{E}$ is both a ring and a vector lattice — see Proposition 3.10. Let then $f \in \mathcal{C}$ and $\epsilon > 0$ be given. For every pair $s \neq t$ in S find a function $\psi_{s,t} \in \mathcal{E}$ with $\psi_{s,t}(s) \neq \psi_{s,t}(t)$. For any $s \in S$ let $\psi_s \in \mathcal{E}$ be a function with $\psi_s(s) = 1$. This is impossible only in case (ii) and when $s = z$, in which case we set $\psi_s = 0$ and observe that $f(s) = f(z) = 0$. Then define $\phi_{s,t}$ by

$$\phi_{s,t}(\tau) = f(s)\psi_s(\tau) + \left(\frac{f(t)\psi_t(\tau) - f(s)\psi_s(\tau)}{\psi_{s,t}(t) - \psi_{s,t}(s)}\right) \cdot (\psi_{s,t}(\tau) - \psi_{s,t}(s)) .$$

This function belongs to $\mathcal{E}$ and takes at s and t the same values as f. Fix $t \in S$ and consider the sets

$$U_s^t = [\phi_{s,t} > f - \epsilon] , \qquad s \in S.$$

They are open, and they cover S. Indeed, the point $s \in S$ belongs to U_s^t. Since S is compact, there is a finite subcover $\{[\phi_{s_i,t} : 1 \leq i \leq n\}$. Set $\phi^t = \bigvee_{i=1}^n \phi_{s_i,t}$. This function belongs to $\mathcal{E}$, is everywhere bigger than $f - \epsilon$, and coincides with f at t. Next consider the open cover $\{[\phi^t < f + \epsilon] : t \in S\}$. It has a finite subcover $\{[\phi^{t_i} < f] : 1 \leq i \leq k\}$, and the function $\phi = \bigwedge_{i=1}^k \phi^{t_i} \in \mathcal{E}$ is clearly uniformly as close as ϵ to f. ∎

Corollary 3.12 *Let $\mathcal{E}$ be a ring of bounded functions on some set. Then the uniform closure $\overline{\mathcal{E}}$ of $\mathcal{E}$ is both a ring and a vector lattice closed under chopping. Moreover, $\overline{\mathcal{E}}$ is closed under composition with continuous functions that vanish at zero; that is to say, if $\overline{\phi} \in \overline{\mathcal{E}}$ and $f : \mathbb{R} \to \mathbb{R}$ is continuous and vanishes at zero, then $f \circ \overline{\phi} \in \overline{\mathcal{E}}$.*

Proof. From Proposition 3.10 we know that $\overline{\mathcal{E}}$ is a ring and vector lattice closed under chopping; only the last statement remains to be proved. Let $\epsilon > 0$ be given and let $M = \|\overline{\phi}\|_u + 1$. Since f is uniformly continuous on the interval S, there is a $\delta > 0$ such that $|s - t| < \delta$ implies $|f(s) - f(t)| < \epsilon \quad \forall s, t \in S$. We may assume that $\delta < 1$. There is a function $\phi \in \mathcal{E}$ with $|\overline{\phi}(x) - \phi(x)| < \delta \quad \forall x$. Since the polynomials with zero constant term form a ring separating the points of S, and since f vanishes on the common zero set $\{0\}$ of them, there is one of them, say P, with $|f(t) - P(t)| < \epsilon$ for all $t \in [-M, M]$ (see Theorem 3.11.) The inequalities

$$|f(\overline{\phi}(x)) - P(\phi(x))| \le |f(\overline{\phi}(x)) - f(\phi(x))| + |f(\phi(x)) - P(\phi(x))| < \epsilon + \epsilon = 2\epsilon \ ,$$

true for all x, show that $f \circ \overline{\phi} \in \overline{\mathcal{E}}$, inasmuch as $P \circ \phi \in \mathcal{E}$. ∎

Corollary 3.13 *Let $\mathcal{E}$ be a ring or vector lattice closed under chopping of bounded functions on some set S, and A_0 a subset of S. A function $f : A_0 \to \mathbb{R}$ can be approximated uniformly on A_0 by functions from $\mathcal{E}$ or $\overline{\mathcal{E}}$ if and only if it is the restriction to A_0 of a function in $\overline{\mathcal{E}}$.*

Proof. The condition is clearly sufficient. For the necessity suppose then that the sequence (ϕ_n) in $\mathcal{E}$ or $\overline{\mathcal{E}}$ converges uniformly on A_0 to f. By taking a subsequence we may assume that $\|f - \phi_n\|_u \le 2^{-n-1}$. Then $\|\phi_{n+1} - \phi_n\|_u \le 2^{-n}$, and the function

$$\overline{\psi}_n \stackrel{\text{def}}{=} -2^{-n} \vee \big(\phi_{n+1} - \phi_n\big) \wedge 2^{-n} \ ,$$

which by Proposition 3.10 and Exercise 2.14 belongs to $\overline{\mathcal{E}}$, equals $\phi_{n+1} - \phi_n$ on A_0. Therefore the sum

$$\overline{\phi} \stackrel{\text{def}}{=} \phi_1 + \sum_{n=1}^{\infty} \overline{\psi}_n$$

converges uniformly, with limit in $\overline{\mathcal{E}}$ by Exercise 3.9. Clearly $f = \overline{\phi}$ on A_0. ∎

Supplements and Additional Exercises

Exercise 3.14 Give examples to show that the conclusion of Theorem 3.11 may fail if S is not compact or if $\mathcal{E}$ does not separate points.

Exercise 3.15* (i) A continuous function on the line can be approximated uniformly by polynomials, on every compact set. (This is Weierstraß' original theorem.)
(ii) Find a continuous function on the line that cannot be approximated uniformly by polynomials. (iii) However, every continuous function ϕ on the line can be approximated pointwise by a sequence of polynomials. If ϕ vanishes at zero then polynomials with zero constant term will do.

In Exercises 3.16–3.20 $\mathcal{E}$ is a ring of bounded functions—for a vector lattice closed under chopping their conclusions are generally trivially true.

Exercise 3.16* (i) Every positive $\overline{\phi} \in \overline{\mathcal{E}}$ is the uniform limit of a sequence of positive functions in $\mathcal{E}$. (ii) $\overline{\mathcal{E}}$ contains the constants if and only if $\mathcal{E}$ contains a positive function ϕ that is bounded away from zero.

Exercise 3.17* A continuous function ψ on a compact interval I can be approximated there uniformly by an *increasing* sequence of polynomials. Therefore if $\overline{\mathcal{E}}$ contains the constants then every $\overline{\phi} \in \overline{\mathcal{E}}$ is the uniform limit of an increasing and also of a decreasing sequence in $\mathcal{E}$.

Exercise 3.18* Let $M > 0$. The function $t \mapsto t \wedge 1$ can be approximated uniformly on $[0, M]$ by an increasing sequence of positive polynomials that vanish at zero. For every positive $\phi \in \mathcal{E}$ there is therefore a sequence of positive functions in $\mathcal{E}$ that increases uniformly to $\phi \wedge 1$.

Exercise 3.19* Fix $M > 0$ and $\epsilon > 0$. (i) The function $t \mapsto \psi(t) \stackrel{\text{def}}{=} (t-\epsilon)_+ = t - (t \wedge \epsilon)$ is the uniform limit of both an increasing and a decreasing sequence of polynomials with zero constant term. For every $\phi \in \mathcal{E}$ the function $(\phi - \epsilon)_+ = \phi - (\phi \wedge \epsilon)$ is therefore the uniform limit of both an increasing and a decreasing sequence in $\mathcal{E}$.
(ii) The set $\overline{\mathcal{E}}_{00}$ of $\mathcal{E}$–confined functions in $\overline{\mathcal{E}}$ is a self–confining lattice ring uniformly dense in $\overline{\mathcal{E}}$.

Exercise 3.20 Let $0 < \alpha, M, \epsilon < \infty$. There is a polynomial $P(t)$ with zero constant term such that for all $t \in [0, M]$

$$0 \le P(t) \le t^\alpha \quad \text{and} \quad \left|t^\alpha - P(t)\right| \le \epsilon\,.$$

[Hint: For $\alpha \le 1$ approximate $\alpha t^{\alpha-1} \wedge K$ by a polynomial and let P be its antiderivative.]

Exercise 3.21 Show that the finite linear combinations of functions of the form $t \mapsto \phi_\alpha(t) = e^{-\alpha t}$, $t \in [0, \infty)$, $\alpha > 0$, are uniformly dense in the set $C_0[[0, \infty)]$ of continuous functions on $[0, \infty)$ that vanish at infinity. Deduce that the Laplace transform is one–to–one — recall that the Laplace transform $\mathcal{L}f$ of a continuous bounded function $f : [0, \infty) \to \mathbb{R}$ is defined by

$$\mathcal{L}f(\alpha) = \int_0^\infty f(t) e^{-\alpha t}\, dt,\ \alpha > 0\,.$$

Exercise 3.22 Let S be a compact space and $\mathcal{E}$ a *complex* ring of complex–valued continuous functions on S that separates the points of S and is closed under taking complex conjugates.
(i) Show that every complex–valued continuous function f on S that vanishes where every function of $\mathcal{E}$ vanishes is the uniform limit of a sequence in $\mathcal{E}$.
(ii) Show that the conclusion of (i) fails if S is not compact **or** if $\mathcal{E}$ does not separate the points **or** if $\mathcal{E}$ is not closed under taking complex conjugates.

Another Version of the Stone–Weierstraß Theorem requires no assumption of compactness, nor does the approximating class $\mathcal{E}$ need to separate the points. We shall not refer to the remaining material of this subsection in the main body, so it may be skipped on first reading or altogether—it is meant for the compulsive aesthetic.

The data are a set S and a ring or vector lattice closed under chopping, of bounded real–valued functions on S. In order to discuss this general theorem we must introduce the $\mathcal{E}$*–uniformity* on S. To this end note that for every $\phi \in \mathcal{E}$

$$d_\phi(x, y) = |\phi(x) - \phi(y)|$$

defines a pseudometric d_ϕ on S: clearly, every d_ϕ is symmetric, $d_\phi(x, y) = d_\phi(y, x)$, and the triangle inequality $d_\phi(x, z) \le d_\phi(x, y) + d_\phi(y, z)$ obtains. The collection

$$\mathcal{D}[\mathcal{E}] = \{d_\phi : \phi \in \mathcal{E}\}$$

of pseudometrics is called the $\mathcal{E}$**–*uniformity*** on S. A function f on S with values in some metric space with metric d is called $\mathcal{E}$**–*uniformly continuous*** if for every $\epsilon > 0$ there are a $\delta > 0$ and finitely many pseudometrics $d_{\phi_1}, \ldots, d_{\phi_n}$ in $\mathcal{D}[\mathcal{E}]$ such that

$$d_{\phi_1}(x,y) < \delta, \ \ldots, \ d_{\phi_n}(x,y) < \delta \ \text{ imply } \ d(f(x), f(y)) < \epsilon \, .$$

Exercise 3.23 The uniform limit of a sequence of $\mathcal{E}$–uniformly continuous functions is $\mathcal{E}$–uniformly continuous.

Exercise 3.24 Let E be another space equipped with a collection $\mathcal{D}$ of pseudometrics. Define the notion of uniform continuity for a function $f : S \to E$. Next let $\mathcal{F}$ be a collection of uniformly continuous functions from S to E. Define the uniform closure $\overline{\mathcal{F}}^u$ of $\mathcal{F}$ and show that it consists again of uniformly continuous functions.

Theorem 3.25 (The General Stone–Weierstrass Theorem) *Let S be a set and $\mathcal{E}$ a ring, or a vector lattice closed under chopping, of bounded real–valued functions on S. Every real–valued $\mathcal{E}$–uniformly continuous function f on S is the sum of a constant and a function in the uniform closure $\overline{\mathcal{E}}$ of $\mathcal{E}$.*

Proof. Consider the product of compact intervals

$$\Pi = \prod_{\phi \in \mathcal{E}} [-\|\phi\|_u, \|\phi\|_u] \, .$$

A typical element of Π is an "$\mathcal{E}$–tuple" $\{x_\phi : \phi \in \mathcal{E}\}$ with $-\|\phi\|_u \le x_\phi \le \|\phi\|_u \ \forall \phi \in \mathcal{E}$. Under the product topology, Π is a compact Hausdorff space. It has a natural uniformity, given by the collection of "coordinate pseudometrics"

$$\widehat{d}_\psi(\{x_\phi : \phi \in \mathcal{E}\}, \{y_\phi : \phi \in \mathcal{E}\}) = |x_\psi - y_\psi| \ ,$$

one for every $\psi \in \mathcal{E}$. The topology associated with this uniformity is clearly nothing but the product topology. There is a natural map $j : S \to \Pi$, to wit,

$$S \ni x \mapsto \{\phi(x) : \phi \in \mathcal{E}\} \, .$$

Let $\overline{S}$ denote the closure of $j(S)$ in Π. This is again a compact space (it is called the ***Hausdorff completion*** of S). Let $x, y \in S$ be two points such that $j(x) = j(y)$. Then $d_\phi(x,y) = 0$ for all $d_\phi \in \mathcal{D}[\mathcal{E}]$, and consequently $f(x) = f(y)$ for all uniformly continuous functions f on S. For every such f there is thus a unique function f' on $j(S)$ such that

$$f = f' \circ j \, .$$

It is easy to see that f' is uniformly continuous on $j(S)$. There is therefore a unique uniformly continuous extension $\widehat{f}$ of f' to $\overline{S}$. Let $\widehat{\mathcal{E}}$ be the collection of functions $\widehat{\phi}:\overline{S}\to\mathbb{R}$, constructed from $\phi\in\mathcal{E}$ in the way $\widehat{f}$ is constructed from f. This is clearly a ring or a vector lattice closed under chopping as the case may be, and it separates the points of $\overline{S}$. The Stone–Weierstraß Theorem 3.11 applies. Suppose first there is a point $z\in\overline{S}$ at which the functions $\widehat{\phi}$ all vanish. Then $\widehat{f_0}=\widehat{f}-\widehat{f}(z)$ is the uniform limit of a sequence $(\widehat{\phi}_n)$ in $\widehat{\mathcal{E}}$, and consequently $f=\widehat{f}\circ j$ is the sum of the number $\widehat{f}(z)$ and the uniform limit of the sequence $(\widehat{\phi}_n\circ j)$ in $\mathcal{E}$. If there is no common zero z of $\widehat{\mathcal{E}}$, the argument is even simpler. ▮

Corollary 3.26 *A real-valued $\mathcal{E}$-uniformly continuous function is bounded. It is the sum of a constant and a function in $\overline{\mathcal{E}}^u$.*

Exercise 3.27 In the situation of Exercise 3.24, show that any $\mathcal{E}$-uniformly continuous function f from S to E has precompact range.

Exercise 3.28 Let $\mathcal{E}$ be a *self-confining* ring or vector lattice closed under chopping, of bounded functions on some set S. We know from Exercise 3.19 that $\overline{\mathcal{E}}_{00}^u$ is a lattice ring. Show that every $\overline{\phi}\in\overline{\mathcal{E}}_{00}$ is the uniform limit of some sequence (ϕ_n) in $\mathcal{E}$ all of whose members are confined by the same function $\psi\in\mathcal{E}$. In other words, $\overline{\phi}$ is the "confined uniform limit" of a sequence in $\mathcal{E}$.

I.4 The Riemann Integral

The most frequently encountered construction of the Riemann integral is this: First of all, one attempts to integrate only functions f that are bounded and vanish outside some bounded interval $(a,b]$. Let us call $\mathcal{F}^\natural$ their collection. Given $f\in\mathcal{F}^\natural$, partitions

$$\mathcal{P}=\{a=x_0<x_1<x_2<\ldots<x_n=b\}$$

are invoked to define the upper and lower Riemann sums

$$\begin{aligned} U(f;\mathcal{P}) &= \sum_{i=1}^{n}\left(\sup_{x_{i-1}<x\le x_i} f(x)\right)\times(x_i-x_{i-1}) \qquad \text{and} \\ L(f;\mathcal{P}) &= \sum_{i=1}^{n}\left(\inf_{x_{i-1}<x\le x_i} f(x)\right)\times(x_i-x_{i-1})\,. \end{aligned} \tag{4.1}$$

The prospective integral of f should clearly lie between these two numbers, and thus between the upper and lower Riemann integrals of f, which are defined as

$$\int^{\natural} f=\inf\, U(f;\mathcal{P}) \quad\text{and}\quad \int_{\natural} f=\sup\, L(f;\mathcal{P})\;, \tag{4.2}$$

respectively. The infimum and supremum are taken over all possible partitions $\mathcal{P}$ of the interval $(a,b]$. There may be functions f for which $\int^{\natural} f \neq \int_{\natural} f$, in which case there is doubt what the integral of f should be. If, however, these two numbers agree then there is no doubt:

Definition 4.1 *The function f is **Riemann integrable** if $\int_{\natural} f = \int^{\natural} f$, and the common value is the integral of f, denoted variously by*

$$\int f(x)\,dx \quad \textit{or} \quad \int f(x)\,\lambda(dx) \quad \textit{or} \quad \int f\,d\lambda \quad \textit{or simply by} \quad \int f\,.$$

The intuition behind this is, of course, that $U(f;\mathcal{P})$ is the integral of a step function $\overline{\phi}$ majorizing f and $L(f;\mathcal{P})$ that of a step function $\underline{\phi}$ below f. Namely,

$$\overline{\phi}(x) = \begin{cases} \sup\{f(x) : x_{i-1} < x \le x_i\} & \text{for } x \in (x_{i-1}, x_i]\,, i = 1 \ldots n, \\ 0 & \text{elsewhere} \end{cases}$$

and

$$\underline{\phi}(x) = \begin{cases} \inf\{f(x) : x_{i-1} < x \le x_i\} & \text{for } x \in (x_{i-1}, x_i]\,, i = 1 \ldots n, \\ 0 & \text{elsewhere.} \end{cases}$$

Since $\underline{\phi} \le f \le \overline{\phi}$, the integrals of these three functions should satisfy the same order relation, to wit,

$$\int \underline{\phi} \le \int f \le \int \overline{\phi}\,.$$

In other words, the requirement that the prospective integral respect the order —the larger the integrand the larger its integral—forces the conclusion that the maximal and minimal possible values of $\int f$ are

$$\begin{aligned} \int^{\natural} f &= \inf\{\int \phi : \phi \text{ a step function } \ge f\} \quad \text{and} \\ \int_{\natural} f &= \sup\{\int \phi : \phi \text{ a step function } \le f\}, \text{ respectively.} \end{aligned} \tag{4.3}$$

If the two numbers on the left in (4.3) disagree then there is doubt how $\int f$ should be defined; we sadly discard f as non–integrable. If they agree there is no doubt; we call the function f integrable and define the integral as the common value of $\int_{\natural} f = \int^{\natural} f$. We shall denote by $\mathcal{L}^{\natural}[\mathbb{R}]$ or $\mathcal{L}^{\natural}$ the collection of Riemann integrable functions.

To make all this precise we should define what a step function on the line is.

Definition 4.2 *A function $\phi : \mathbb{R} \to \mathbb{R}$ is called a **step function on the line** if there exists a partition $\mathcal{P} = \{x_0 < x_1 < x_2 < \cdots < x_n\}$ of the line such that ϕ is constant on the intervals $(x_{i-1}, x_i]$, $i = 1, \ldots, n$, and zero outside their union*

$(x_0, x_n]$. *Any such partition — and there are many in general — will be called a* ***partition for*** ϕ. *The collection of all step functions on the line is denoted by* $\mathcal{E}[\mathbb{R}]$.

Proposition 4.3 *The step functions* $\mathcal{E}[\mathbb{R}]$ *form a lattice ring of bounded functions. For every step function* ϕ *there is another step function* ψ *confining* ϕ, *meaning that* ϕ *vanishes outside the set* $[\psi \geq 1]$. *In the terms of Definition 2.16,* $\mathcal{E}[\mathbb{R}]$ *is self–confining.*

The ***elementary integral*** $\phi \mapsto \int \phi$ *is a linear and positive map from* $\mathcal{E}[\mathbb{R}]$ *to the reals — to say that* $\phi \mapsto \int \phi$ *is* ***positive*** *means that* $\phi \geq 0$ *implies* $\int \phi \geq 0$ *and has the evident consequence that* $\int$ *is increasing:* $\phi \leq \phi'$ *implies* $\int \phi \leq \int \phi'$.

Proof. Let ϕ, ϕ' be step functions and $\mathcal{P}, \mathcal{P}'$ partitions for them, and let $r, r' \in \mathbb{R}$. The union of $\mathcal{P}$ and $\mathcal{P}'$ is a partition $\mathcal{Q}$ for both ϕ and ϕ'. It is obvious that $r\phi + r'\phi'$, $\phi \cdot \phi'$, $\phi \wedge \phi'$, $\phi \vee \phi'$, and $\phi \wedge 1$ are constant on the intervals of $\mathcal{Q}$ and zero outside their union and thus are step functions. For the next statement, let ψ be the (indicator function of the) set $[\phi \neq 0]$, which belongs to $\mathcal{E}[\mathbb{R}]$. The last statement is left to the reader. ∎

Remark 4.4 We chose half–open intervals of the form $(a, b]$ for the intervals of constancy of a step function – this is the usual convention. It is possible to choose intervals of the form $[a, b)$ instead. One must, however be consistent. Allowing intervals of both or other forms leads to very messy arguments in Proposition 4.3.
The Definitions (4.2) and (4.3) of the upper and lower integrals look different but produce the same numbers (see Exercise 4.9 and Exercise 4.11). The first of them obscures the functionality of the definition of the integral through the requirement of preservation of order by insisting that the estimating lower and upper step functions are optimal given their partition; this is really a side issue concerning efficiency of computation.

Let us collect in one spot the pertinent properties of the upper integral:

Proposition 4.5 *The Riemann upper integral is*

(i) positive–homogeneous: $\int^{\natural} r \cdot f = r \cdot \int^{\natural} f \qquad \forall r \in \mathbb{R}_+, \forall f \in \mathcal{F}^{\natural}$;

(ii) subadditive: $\int^{\natural} (f + f') \leq \int^{\natural} f + \int^{\natural} f' \qquad \forall f, f' \in \mathcal{F}^{\natural}$;

(iii) increasing: $0 \leq f \leq f' \implies \int^{\natural} f \leq \int^{\natural} f' \qquad \forall f, f' \in \mathcal{F}^{\natural}$;

(iv) and majorizes the integral $|\int \phi| \leq \int^{\natural} |\phi| \qquad \forall \phi \in \mathcal{E}[\mathbb{R}]$.

Proof. (iii) and (iv) are obvious. (i): Let $\int^{\natural} f < s$. Then there is a $\phi \in \mathcal{E}$ with $f \leq \phi$ and $\int \phi < s$. Then $r\phi \in \mathcal{E}$ majorizes rf and has integral $r \int \phi < rs$. Thus $\int^{\natural} rf < rs$, and consequently $\int^{\natural} rf \leq r \int^{\natural} f$ — see Exercise 2.3 on p. 4. The reverse inequality follows from applying the previous one to f/r: $\int^{\natural} f = \int^{\natural} r \cdot f/r \leq r \cdot \int^{\natural} f/r$; replacing r by $1/r$ yields $r \int^{\natural} f \leq \int^{\natural} rf$ as desired. In case $r = 0$ there is nothing to prove.

(ii): Understanding well the very simple argument that leads from the additivity of the integral on the elementary functions to the subadditivity of the upper integral on all functions will take any mystery out of the future development.

Suppose s is a real number strictly exceeding $\int^{\natural} f + \int^{\natural} f'$. Then there exist step functions $\phi \geq f$ and $\phi' \geq f'$ with $\int \phi + \int \phi' < s$. Now $\phi + \phi'$ is a step function majorizing $f + f'$, and the additivity of $\int$ on $\mathcal{E}$ gives

$$\int^{\natural} (f + f') \leq \int (\phi + \phi') = \int \phi + \int \phi' < s .$$

Hence $\int^{\natural} (f + f') \leq \int^{\natural} f + \int^{\natural} f'$, as claimed (use Exercise 2.3). ∎

Additional Exercises

Exercise 4.6* (i) A set A belongs to $\mathcal{E}[\mathbb{R}]$ if and only if it is the finite union of half–open intervals of the form $(a, b]$, i.e., $A = \bigcup_{i=1}^{I} (a_i, b_i]$. Of the many representations of A in this form there is one where the intervals $(a_i, b_i]$ are disjoint and non–adjacent, and this representation is unique. We call it the *minimal representation* of A.

(ii) The collection $\mathcal{A}[\mathbb{R}]$ of sets in $\mathcal{E}[\mathbb{R}]$ is closed under taking finite unions, finite differences, and relative complements. (A collections of subsets of some set that has these three closure properties is called a *ring of subsets* — this notion is discussed in detail in Proposition II.2.13.)

(iii) $\mathcal{E}[\mathbb{R}]$ consists exactly of the linear combinations of functions in $\mathcal{A}[\mathbb{R}]$ (Convention 2.6 on p. 6) — we say $\mathcal{A}[\mathbb{R}]$ linearly generates $\mathcal{E}[\mathbb{R}]$ —, and the elementary integral is the unique extension by linearity of its restriction to $\mathcal{A}[\mathbb{R}]$. This restriction is denoted by λ and is called *Lebesgue measure* on $\mathcal{A}[\mathbb{R}]$.

Exercise 4.7 A function $\phi : \mathbb{R} \to \mathbb{R}$ belongs to $\mathcal{E}[\mathbb{R}]$ if and only if it is left–continuous, has only finitely many values, and vanishes outside a bounded set.

Exercise 4.8 Define the notion of a partition of the plane $\mathbb{R}^2$ into half–open rectangles $(a, b] \times (c, d]$. Use this to define step functions $\mathcal{E}[\mathbb{R}^2]$ and define the elementary integral $\int : \mathcal{E}[\mathbb{R}^2] \to \mathbb{R}$. Redo Exercise 4.6 for this situation, and develop the Riemann integral in the plane this way.

Exercise 4.9 Prove that the two Definitions (4.1) and (4.3) of $\int^{\natural} f$ and $\int_{\natural} f$ do, indeed, coincide. Then show that $\int_{\natural} f = -\int^{\natural} (-f)$.

Exercise 4.10 (i) Define the notion of negative–homogeneity and of superadditivity. (ii) Show that $\int_{\natural}$ is positive–homogenous, increasing, and superadditive.

(iii) Show that neither $\int^{\natural}$ nor $\int_{\natural}$ are negative–homogeneous.

Exercise 4.11 In many Calculus textbooks, the maxima and minima in the definition (4.1) of the upper and lower Riemann sums are extended over the closed intervals $[x_{i-1}, x_i]$ instead of the half–open ones, thus "counting the values of f at the endpoints x_{i-1} twice." Prove that this does not affect the values of $\int^{\natural} f$ and $\int_{\natural} f$ in (4.1).

Exercise 4.12 Compute $\int^{\natural} f$ for $f(x) = (0, 1](x) \cdot e^x$.

I.5 An Integrability Criterion

Now that the Riemann integrable functions have been identified, the next task is, as the reader remembers from the Calculus, to establish the properties of the integral: linearity, positivity, limit theorems, etc. It will become clear in Section 6 that they are entirely due to the structure of $\mathcal{E}$ and to the properties of the upper integral $\int^{\natural}$ listed in Propositions 4.3 and 4.5. The task will be facilitated greatly by the integrability criterion of the present section.

The definition of integrability given in the previous section is intuitive enough, reflecting as it does the order properties we expect from the integral. Evidently, an integrable function f is rather close to being a step function, since it can be squeezed arbitrarily closely by the latter. The following proposition will lead to a precise formulation of this somewhat vague statement.

Proposition 5.1 *Let f be a function in $\mathcal{F}^{\natural}$. The following are equivalent:*
(i) f is Riemann integrable — i.e., $\infty < \int^{\natural} f = \int_{\natural} f < +\infty$;
(ii) For every $\epsilon > 0$ there is a step function ϕ such that $\int^{\natural} |f - \phi| < \epsilon$;
(iii) There exists a sequence (ϕ_n) of step functions with

$$\lim_{n\to\infty} \int^{\natural} |f - \phi_n| = 0 . \tag{5.1}$$

In this case, for any sequence (ϕ_n) of step functions satisfying (5.1),

$$\int f = \lim_{n\to\infty} \int \phi_n .$$

Proof. (i)$\Rightarrow$(ii): Given $\epsilon > 0$, one can find step functions $\underline{\phi}, \overline{\phi}$ with $\underline{\phi} \le f \le \overline{\phi}$ and $\int \overline{\phi} - \int \underline{\phi} < \epsilon$. Then

$$|f - \underline{\phi}| \;=\; f - \underline{\phi} \;\le\; \overline{\phi} - \underline{\phi} .$$

Since $\int(\overline{\phi} - \underline{\phi}) = \int \overline{\phi} - \int \underline{\phi} < \epsilon$, we have $\int^{\natural} |f - \underline{\phi}| < \epsilon$ by the very definition of $\int^{\natural}$.
(ii)$\Rightarrow$(i): Assume that $\int^{\natural} |f - \phi|$ can be made arbitrarily small by the choice of a step function ϕ. Given $\epsilon > 0$, we can then find one such that $\int^{\natural} |f - \phi| < \epsilon/2$. By the definition of the upper integral, there is a step function

$$\rho \ge |f - \phi| \quad \text{with} \quad \int \rho < \epsilon/2 .$$

Then $\underline{\phi} = \phi - \rho \le f$, $\overline{\phi} = \phi + \rho \ge f$, and

$$\int \underline{\phi} \le \int_{\natural} f \le \int^{\natural} f \le \int \overline{\phi} = \int \underline{\phi} + 2 \cdot \int \rho < \int \underline{\phi} + \epsilon .$$

Thus $\int_{\natural} f$ and $\int^{\natural} f$ cannot differ by more than ϵ. Since $\epsilon > 0$ was arbitrary the numbers $\int_{\natural} f$ and $\int^{\natural} f$ agree, and f is integrable as claimed.
(ii) and (iii) are clearly but restatements of each other. The last claim follows from the following consequence of the subadditivity, which is left for the reader to establish:

$$\left| \int^{\natural} f - \int^{\natural} f' \right| \leq \int^{\natural} |f - f'| \ . \tag{5.2}$$

Indeed, then

$$\left| \int f - \int \phi_n \right| = \left| \int^{\natural} f - \int^{\natural} \phi_n \right| \leq \int^{\natural} |f - \phi_n| \xrightarrow[n\to\infty]{} 0 \ . \qquad \blacksquare$$

Let us attach to every function $f \in \mathcal{F}^{\natural}$ a number $\|f\|^{\natural}$ reflecting its size by

$$\|f\|^{\natural} = \int^{\natural} |f| \ ,$$

and use it to measure the distance of two functions f, ϕ from each other *via*

$$\|f - \phi\|^{\natural} \ .$$

Conditions (ii)–(iii) express the idea that for f to be Riemann integrable it must be approximable by step functions in the sense that the distance of f from the collection $\mathcal{E}$ of step functions, i.e., the infimum of the distances

$$\|f - \phi\|^{\natural} \ , \qquad \phi \in \mathcal{E},$$

is zero. The measure $\|f\|^{\natural}$ of the size of f is different from $\|f\|_u$ and from other ways of attaching a size to a function f, but it has the virtue of being adapted to the problem at hand. What matters in the present context is not the maximal value of $|f|$ but the infimum $\|f\|^{\natural}$ of the integrals of step functions exceeding $|f|$. This is common practice; for example, an investor interested in long–term investment will measure the merit of a stock's performance by its growth, one whose interest is in cash flow by its annual yield. In fact, it is a most fruitful device in the modern analysis of functions — for instance in the theory of differential equations — to start by assigning functions a size that reflects their utility for the problem at hand. The utility of $\| \ \|^{\natural}$ for integration lies in that it majorizes the elementary integral:

$$\left| \int \phi \right| \leq \|\phi\|^{\natural} \qquad \forall \phi \in \mathcal{E} \ .$$

Definition 5.2 *The functional*[2] $\| \ \|^\natural$ *on* $\mathcal{F}^\natural$ *is called the* ***Jordan mean****, after the French mathematician Jordan who first employed it. A sequence* (f_n) *of functions in* $\mathcal{F}^\natural$ *is said to converge in Jordan mean to the function* f *if*

$$\lim_{n\to\infty} \|f - f_n\|^\natural = 0 .$$

The integrability criterion 5.1 at the beginning of this section may then be rephrased thus: The function f is Riemann integrable precisely if it can be approximated arbitrarily closely by step functions $\phi \in \mathcal{E}$, in Jordan mean; and if (ϕ_n) is any sequence of step functions so approximating f, then $\int f = \lim_{n\to\infty} \int \phi_n$. The reader having some background in topology sees that this means but the following: $\mathcal{L}^\natural$ is the closure of $\mathcal{E}$ in $\mathcal{F}^\natural$ with respect to the pseudo–metric $\operatorname{dist}(f,\phi) = \|f-\phi\|^\natural$, and the Riemann integral is the unique continuous extension of the elementary integral.

The properties of the Riemann integral are all consequences of the structure of $\mathcal{E}$ and of the following four properties of $\| \ \|^\natural$, which are evident from Proposition 4.5 on p. 19.

Proposition 5.3 *The Jordan mean* $\| \ \|^\natural$ *is*

subadditive:	$\|f+f'\|^\natural \le \|f\|^\natural + \|f'\|^\natural$;
absolute–homogeneous:	$\|r\cdot f\|^\natural = \lvert r\rvert\cdot\|f\|^\natural$;
and solid:	$\lvert f\rvert \le \lvert g\rvert \implies \|f\|^\natural \le \|g\|^\natural$.
It majorizes the integral:	$\lvert \int f\rvert \le \int \lvert f\rvert \le \|f\|^\natural \quad \forall f \in \mathcal{L}^\natural$.

I.6 The Permanence Properties

The reader remembers from the Calculus that one does, in practice, rarely check a given function f for integrability by applying the original definition or even a criterion such as Proposition 5.1. Rather, one tries to see that f can be obtained from functions known to be integrable by algebraic, order–, or limit–operations, or by other explicit constructions that are known to preserve integrability. The list of such operations and constructions constitutes the "*permanence properties of the integral.*" The permanence properties of a notion are, as a rule, closer to its nuts–and–bolts applications than is its definition[3].

---

[2] A functional is a function whose arguments are functions or elements of some "abstract vector space."

[3] It is possible to imagine a student who, without understanding the notion of a derivative, coasts successfully through his Calculus course by merely memorizing and applying its permanence properties: sum rule, Leibniz' rule, chain rule etc., plus the derivatives of a few functions

Exercise 6.1 Recall the definitions of continuity and of differentiability. How do you know that the function $x \mapsto e^{-|x|}$ is continuous? How do you know that a power series yields a continuous function? How do you know that $x \mapsto \arctan(x^2 + x^3)$ is differentiable? List the permanence properties of continuity and of differentiability.

Here is a simple permanence property of the Riemann integral:

Theorem 6.2 *If the sequence (f_n) of Riemann integrable functions converges in Jordan mean to f then f is Riemann integrable and*

$$\int f = \lim_{n\to\infty} \int f_n .$$

Proof. For every $n \in \mathbb{N}$ find $\phi_n \in \mathcal{E}$ with $\|f_n - \phi_n\|^\natural < 1/n$. By the subadditivity of $\| \ \|^\natural$,

$$\|f - \phi_n\|^\natural \le \|f - f_n\|^\natural + \|f_n - \phi_n\|^\natural < \|f - f_n\|^\natural + 1/n \xrightarrow[n\to\infty]{} 0 ,$$

so f is integrable. Since $\| \ \|^\natural$ majorizes the integral,

$$|\int f - \int f_n| \le \int |f - f_n| \le \|f - f_n\|^\natural \xrightarrow[n\to\infty]{} 0 .$$

Therefore $\int f = \lim \int \phi_n = \lim \int f_n$ as claimed. ∎

The Algebraic and Order Permanence Properties of the Riemann integral get our attention first. There are seven listed in the following theorem. The proof is the same for every one of them and makes use only of the structure of the step functions as a lattice ring of bounded functions (Proposition 4.3) and of the fact that $\| \ \|^\natural$ is subadditive, absolute–homogeneous, and solid (Proposition 5.3). We carry it out in detail, since it applies literally also to the Daniell mean $\| \ \|^*$ of Chapter II.

Theorem 6.3 *Let f, f' be Riemann integrable functions and let r be a real number. Then $f + f'$, rf, $f \vee f'$, $f \wedge f'$, $|f|$, $f \wedge 1$, and $f \cdot f'$ are Riemann integrable. In other words, the space $\mathcal{L}^\natural(\mathbb{R})$ of Riemann integrable functions is a lattice ring of bounded functions.*

Proof. We start with the sum. For any two functions $\phi, \phi' \in \mathcal{E}$

$$|(f + f') - (\phi + \phi')| \le |f - \phi| + |f' - \phi'|, \quad \text{and so}$$
$$\|(f + f') - (\phi + \phi')\|^\natural \le \|f - \phi\|^\natural + \|f' - \phi'\|^\natural .$$

---

like $x \mapsto x$, $\cos x$, $\sin x$, $\exp x$, $\ln x$. As a matter of fact, I have heard from students who accused their instructor of being "theoretical" when he explained the derivative where simple memorization would do. An accomplished mathematician doing hard–core estimates of singular integrals (whatever that is), will rarely use directly his definition of the Lebesgue integral, which he understands very well, but rather apply its permanence properties.

Since the right–hand side can be made as small as one pleases by the choice of ϕ, ϕ', so can the left–hand side. This says that $f + f'$ is integrable, inasmuch as $\phi + \phi'$ belongs to $\mathcal{E}$. The same argument applies to the other combinations:

$$\begin{aligned}
|(rf) - (r\phi)| &\le |r||f - \phi| \;; \\
|(f \vee f') - (\phi \vee \phi')| &\le |f - \phi| + |f' - \phi'| \;; \\
|(f \wedge f') - (\phi \wedge \phi')| &\le |f - \phi| + |f' - \phi'| \;; \\
||f| - |\phi|| &\le |f - \phi| \;; \\
|f \wedge 1 - \phi \wedge 1| &\le |f - \phi| \;; \\
|(f \cdot f') - (\phi \cdot \phi')| &\le |f||f' - \phi'| + |\phi'||f - \phi| \\
&\le \|f\|_u \cdot |f' - \phi'| + \|\phi'\|_u \cdot |f - \phi| \,.
\end{aligned}$$

These inequalities are true for all functions f, f', ϕ, ϕ' and scalars r. We apply $\| \;\|^\natural$ to them and obtain, using only the subadditivity, absolute–homogeneity, and solidity of $\| \;\|^\natural$,

$$\begin{aligned}
\|(rf) - (r\phi)\|^\natural &\le |r|\|f - \phi\|^\natural \;; \\
\|(f \vee f') - (\phi \vee \phi')\|^\natural &\le \|f - \phi\|^\natural + \|f' - \phi'\|^\natural \;; \\
\|(f \wedge f') - (\phi \wedge \phi')\|^\natural &\le \|f - \phi\|^\natural + \|f' - \phi'\|^\natural \;; \\
\big\||f| - |\phi|\big\|^\natural &\le \|f - \phi\|^\natural \;; \\
\|f \wedge 1 - \phi \wedge 1\|^\natural &\le \|f - \phi\|^\natural \;; \\
\|(f \cdot f') - (\phi \cdot \phi')\|^\natural &\le \|f\|_u \cdot \|f' - \phi'\|^\natural + \|\phi'\|_u \cdot \|f - \phi\|^\natural \;.
\end{aligned}$$

Given an $\epsilon > 0$, we may choose functions $\phi, \phi' \in \mathcal{E}$ so that the right–hand sides are less than ϵ. This is possible because the functions f, f' are integrable and shows that the functions $rf, f \vee f', \ldots$ are integrable as well, inasmuch as the functions $r\phi, \phi \vee \phi', \ldots$ appearing on the left belong to $\mathcal{E}$.
The last case, that of the product, is marginally more complicated than the others. Given $\epsilon > 0$, we choose first $\phi' \in \mathcal{E}$ so that

$$\|f' - \phi'\|^\natural \le \frac{\epsilon}{2(1 + \|f\|_u)} \;,$$

using the fact that the function f is bounded. Then we choose $\phi \in \mathcal{E}$ so that

$$\|f - \phi\|^\natural \le \frac{\epsilon}{2(1 + \|\phi'\|_u)} \;.$$

Then again $\|f \cdot f' - \phi \cdot \phi'\|^\natural \le \epsilon$, showing that $f \cdot f'$ is integrable, inasmuch as the product $\phi \cdot \phi'$ belongs to $\mathcal{E}$. ∎

Exercise 6.4 The ***extended integral*** $\int : \mathcal{L}^\natural(\mathbb{R}) \to \mathbb{R}$ is linear and order preserving.

Theorem 6.3 does not allow us to construct any new integrable functions. For example, if $f, f' \in \mathcal{E}$ then $f \wedge f'$ is integrable all right, but this is nothing new since $\mathcal{E}$ itself is closed under taking finite infima. So Theorem 6.3 by itself is not particularly useful. We have included it mostly for its proof, simple as it is. Subadditive, absolute–homogenous, and solid gauges for the size of functions recur throughout integration theory; the proof above applies literally to all of them and yields the same algebraic and order permanence properties for function classes defined with their aid.

The Permanence Property Concerning Limits is the main source of new Riemann integrable functions:

Theorem 6.5 (The Dominated Uniform Convergence Theorem) *Let (f_n) be a sequence of Riemann integrable functions and assume that*

(i) $f_n \xrightarrow[n\to\infty]{} f$ *uniformly, and*
(ii) $|f_n| \le g$ *for some integrable function g and all indices n.*

Then (f_n) converges to f in Jordan mean, f is integrable, and $\int f = \lim_{n\to\infty} \int f_n$.

Proof. Since every Riemann integrable function is majorized by a step function we may as well assume that the dominating function g itself is one of the latter. By Proposition 4.3, there is a positive step function $\psi \ge 0$ such that g, and then all the f_n, vanish off the set $[\psi \ge 1]$. The domination condition is thus stronger than it looks at first: It forces the f_n all to vanish off a common simple set $[\psi \ge 1]$, or to be *confined* by ψ, as we might say. This implies the *order relation*

$$|f - f_n| \le \psi \cdot \|f - f_n\|_u ,$$

and we can use the *solidity* of $\| \ \|^\natural$ to conclude that

$$\|f - f_n\|^\natural \le \|\psi\|^\natural \cdot \|f - f_n\|_u \xrightarrow[n\to\infty]{} 0 .$$

The claim follows from Theorem 6.2. ∎

Exercise 6.6 The conclusion fails without the domination condition or if (f_n) converges merely pointwise. It fails even when the convergence is uniform and the limit function f is $\mathcal{E}[\mathbb{R}]$–confined.

The only new integrable functions produced from step functions by the theorem are dominated uniform limits of step functions[4]. Yet it yields one of the main results of the Calculus — its proof is left as an exercise:

Corollary 6.7 *A continous function of compact support is Riemann integrable.*

---

[4] There are more: see Exercise 6.16.

Supplements and Additional Exercises

Exercise 6.8 If f is Riemann integrable and if f' is a function that differs from f in but finitely many points then f' is Riemann integrable. Thus any bounded interval is Riemann–integrable, and a step function whose steps are any kind of interval is Riemann integrable. (Note that Theorem 6.3 again does not produce new integrable functions when applied after this result; the step functions over arbitrary intervals already form a ring and vector lattice closed under chopping.)

Exercise 6.9 A piecewise continous function of compact support is Riemann integrable.

Exercise 6.10 It is not usually shown in the Calculus that the infimum and the product of two Riemann integrable functions is again Riemann integrable. Why not?

Exercise 6.11 Let f be Riemann integrable and let $\phi : \mathbb{R} \to \mathbb{R}$ be a continuous function with $\phi(0) = 0$. The composition $\phi \circ f$ is Riemann integrable.

Exercise 6.12 Let $(a, b]$ be a bounded interval on the line and ϕ a continuous function on $(a, b]$ that has a limit at a. The function that equals ϕ on $(a, b]$ and equals zero elsewhere is Riemann integrable.

Exercise 6.13* Let f be a function on the line, and $a < b$. One says ***f is Riemann integrable on*** $(a, b]$ if the function $f \cdot (a, b] = f \cdot 1_{(a,b]}$ is integrable. In this case $\int_a^b f(x)\, dx$ is defined as the integral of that function: $\int_a^b f(x)\, dx \stackrel{\text{def}}{=} \int f(x) \cdot (a, b](x)\, dx$.
(i) Show that $f \cdot (a, b]$ is Riemann integrable if f is continuous on $[a, b]$. Show that this may fail to be true if f is merely continuous on $(a, b]$.
Assume henceforth that f is continuous and bounded on the line.
(ii) Show that $b \mapsto F(b) = \int_a^b f(x)\, dx$ is continuous and differentiable.
(iii) Prove the Fundamental Theorem of Calculus.

Exercise 6.14 Let f be Riemann integrable. Given an $\epsilon > 0$ one can find continuous functions $\underline{\phi}, \overline{\phi}$ of compact support with $\underline{\phi} \le f \le \overline{\phi}$ and $\int (\overline{\phi} - \underline{\phi}) < \epsilon$.

A subset N of the line is called ***Lebesgue–negligible*** if for every $\epsilon > 0$ there is a countable collection of open intervals (a_n, b_n) such that

$$N \subset \bigcup_n (a_n, b_n) \quad \text{and} \quad \sum_n b_n - a_n < \epsilon .$$

N is then also called *a set of Lebesgue measure zero.*

Exercise 6.15 The union of countably many negligible sets is negligible.

Exercise 6.16 A function f is Riemann–integrable if and only if it satisfies the following three properties: (i) it is bounded; (ii) it vanishes outside some bounded set; and (iii) it is continuous except possibly in the points of a Lebesgue–negligible set. (This aesthetically pleasing integrability criterion has little practical value.)

I.7 Seminorms

These are devices to measure the size of functions or of elements of abstract vector spaces. They are most useful tools in Analysis. We have met already two of them: the uniform norm $\| \, \|_u$ and the Jordan mean $\| \, \|^{\natural}$. In fact, it is the theme of these notes that the main ingredient in Lebesgue's integration theory is the clever choice of a seminorm. It is time to give a precise definition of this notion:

Definition 7.1 *Let $\mathcal{F}$ be a vector space. A functional $\| \, \| : \mathcal{F} \to \mathbb{R}_+$ that attaches a real number $\|f\| \geq 0$ to every element f of $\mathcal{F}$ is called a* ***seminorm*** *if it is*

(i) ***subadditive:*** $\|f + f'\| \leq \|f\| + \|f'\| \quad \forall\, f, f' \in \mathcal{F}$

(ii) and ***absolute–homogeneous:*** $\|r \cdot f\| = |r| \cdot \|f\| \quad \forall\, r \in \mathbb{R}\,, \forall f \in \mathcal{F}.$

It is called a ***norm*** *if $\|f\| = 0$ only for $f = 0$. The pair $(\mathcal{F}, \| \, \|)$ is called a semi–normed or normed vector space, accordingly.*
Given a seminorm $\| \, \|$, a sequence (f_n) in $\mathcal{F}$ is said to ***converge*** *to $f \in \mathcal{F}$ (in the sense $\| \, \|$), if $\|f - f_n\| \xrightarrow[n \to \infty]{} 0$. In that case f is the $\| \, \|$*-***limit*** *of (f_n). The statement "(f_n) converges" means that there exists a limit $f \in \mathcal{F}$ to which it converges in the sense of the seminorm.*

The sequence (f_n) in $\mathcal{F}$ is called a ***Cauchy sequence*** *if*

$$\sup_{m,n \geq N} \|f_m - f_n\| \xrightarrow[N \to \infty]{} 0\,.$$

The seminormed vector space $(\mathcal{F}, \| \, \|)$ is ***complete*** *if every Cauchy sequence has a limit in $\mathcal{F}$. A complete normed space is a* ***Banach space.***

The next exercises practice some simple arguments that everyone should do once in his or her life.

Exercise 7.2* The subadditivity (i) is also called the ***triangle inequality.*** It has the consequence that the seminorm is a contractive map from $\mathcal{F}$ to $\mathbb{R}$: for any $f, f' \in \mathcal{F}$

$$\big|\, \|f\| - \|f'\| \,\big| \leq \|f - f'\|\,.$$

Exercise 7.3 A convergent sequence in a seminormed space may have several limits. Describe them. A convergent sequence in a normed space has exactly one limit.

Exercise 7.4 A Cauchy sequence in a seminormed vector space converges if and only if it has a convergent subsequence.

Exercise 7.5 A sequence (f_n) in a seminormed space $(\mathcal{F}, \| \, \|)$ is ***summable*** if the sequence of partial sums $\sum_{\nu=1}^{n} f_\nu$ converges; if so, the limit is denoted by $\sum_{n=1}^{\infty} f_n$. The sequence (f_n) is ***absolutely summable*** if the seminorms $\|f_n\|$ form a summable sequence of reals: $\sum_{n=1}^{\infty} \|f_n\| < \infty$. Show that $(\mathcal{F}, \| \, \|)$ is complete if and only if every absolutely summable sequence is summable.

Exercise 7.6* Let $(\mathcal{L}, \| \ \|)$ be a seminormed space. The set $\mathcal{N} = \{f \in \mathcal{L} : \|f\| = 0\}$ is a vector subspace of $\mathcal{L}$. The quotient $L = \mathcal{L}/\mathcal{N}$ is a normed space under a suitably defined "quotient norm." L is complete if and only if $\mathcal{L}$ is.

The next few exercises are meant to convince the reader that he is already familiar with some (semi)norms and that he is adept at handling them and at contriving new ones.

Exercise 7.7 Whatever the set S, the set $\mathcal{F}_b[S]$ of bounded real–valued functions on S is a vector space under pointwise addition and scalar multiplication, and $\| \ \|_u$ is a norm on $\mathcal{F}_b[S]$. Moreover, $(\mathcal{F}_b[S], \| \ \|_u)$ is complete. The normed spaces $(C_b[\mathbb{R}], \| \ \|_u)$ and $(C_0[\mathbb{R}], \| \ \|_u)$ are complete, while the spaces $(C_b^\infty[\mathbb{R}], \| \ \|_u)$, $(C_{00}[\mathbb{R}], \| \ \|_u)$, and $(C_{00}^\infty[\mathbb{R}], \| \ \|_u)$ are not.

Exercise 7.8 (i) $\| \ \|^\natural$ is a seminorm on $\mathcal{L}^\natural$ but not a norm. (ii) Is $(\mathcal{L}^\natural, \| \ \|^\natural)$ complete?

Exercise 7.9 For $\boldsymbol{x} = (x_1, \ldots, x_n) \in \mathbb{R}^n$ set $\|\boldsymbol{x}\|_1 = \sum_{i=1}^n |x_i|$, $\|\boldsymbol{x}\|_2 = \sqrt{\sum_{i=1}^n x_i^2}$, and $\|\boldsymbol{x}\|_\infty = \sup_{i=1}^n |x_i|$. These are norms on $\mathbb{R}^n$. Note that the spaces $(\mathbb{R}^n, \| \ \|_1)$, $(\mathbb{R}^n, \| \ \|_2)$, and $(\mathbb{R}^n, \| \ \|_\infty)$ all differ as normed spaces, yet have the same underlying vector space $\mathbb{R}^n$ and topology.

Exercise 7.10 For $\boldsymbol{x} \in \mathbb{R}^n$ and $1 \le p < \infty$ set $\|\boldsymbol{x}\|_p = \big(\sum_{i=1}^n |x_i|^p\big)^{1/p}$. $\| \ \|_p$ is a norm on $\mathbb{R}^n$, and $\|\boldsymbol{x}\|_\infty \le \|\boldsymbol{x}\|_q \le \|\boldsymbol{x}\|_p \le \|\boldsymbol{x}\|_1 \le n \cdot \|\boldsymbol{x}\|_\infty$ for $1 \le p \le q < \infty$.

Exercise 7.11 For $\boldsymbol{x} \in \mathbb{R}^{\mathbb{N}}$ and $1 \le p < \infty$ set $\|\boldsymbol{x}\|_p = \big(\sum_{i=1}^\infty |x_i|^p\big)^{1/p}$ and let $\ell^p = \{\boldsymbol{x} : \|\boldsymbol{x}\|_p < \infty\}$. Show that $(\ell^p, \| \ \|_p)$ is a Banach space. Show that $\ell^p \subset \ell^q$ for $1 \le p \le q \le \infty$ and $\ell^p \ne \ell^q$ for $p \ne q$. (These sequence spaces were introduced by M. Riesz in 1906 and are called the "***little*** ℓ^p***–spaces***.")

Definition 7.12 *Let $(\mathcal{F}, \| \ \|)$ be a seminormed vector space and $\mathcal{E}$ a subset of $\mathcal{F}$. The **closure** $\overline{\mathcal{E}}$ of $\mathcal{E}$ is the collection of all $f \in \mathcal{F}$ that can be approximated arbitrarily closely by elements of $\mathcal{E}$, i.e.,*

$$\overline{\mathcal{E}} = \{f \in \mathcal{F} : \quad \forall \epsilon > 0 \quad \exists \phi \in \mathcal{E} \text{ with } \|f - \phi\| < \epsilon\} .$$

*The set $\mathcal{E}$ is **dense** in $\mathcal{F}$ if its closure is all of $\mathcal{F}$. $\mathcal{F}$ is **separable** if it contains a countable dense set.*

Exercise 7.13 $\overline{\overline{\mathcal{E}}} = \overline{\mathcal{E}}$. If $\mathcal{E}$ is a vector subspace of $\mathcal{F}$ then so is $\overline{\mathcal{E}}$.

Exercise 7.14 The space ℓ^p is separable for $1 \le p < \infty$, but ℓ^∞ is not.

Exercise 7.15* The spaces $(C_0[\mathbb{R}], \| \ \|_u)$ and $(\mathcal{L}^\natural[\mathbb{R}], \| \ \|^\natural)$ are separable.

Exercise 7.16* The space $C_{00}[\mathbb{R}]$ of continuous functions of compact support on the line is dense in $(\mathcal{L}^\natural[\mathbb{R}], \| \ \|^\natural)$. So is the space $C_{00}^\infty[\mathbb{R}]$ of infinitely often differentiable functions of compact support.

The next exercise exemplifies the typical use of seminorms in Analysis. An instance of it is Section 5, and the industrious reader should merely transport the arguments there to the present "abstract" setting:

Exercise 7.17* Let $(\mathcal{F}, \| \ \|)$ be a seminormed vector space and $\mathcal{E}$ a vector subspace of $\mathcal{F}$. Let $I : \mathcal{E} \to \mathbb{R}$ be a linear map majorized by the seminorm; that is to say

$|I(\phi)| \le \|\phi\| \quad \forall\, \phi \in \mathcal{E}$. Then there is a unique linear map $\overline{I} : \overline{\mathcal{E}} \to \mathbb{R}$ which is still majorized by the seminorm and agrees with I on $\mathcal{E}$. It is called the ***extension under the seminorm*** $\|\ \|$. Next show that this remains true if the range space $(\mathbb{R}, |\ |)$ of I is replaced by any Banach space $(\mathcal{F}', \|\ \|')$.

Every single one of the seminorms and norms discussed in the previous exercises was solid in the following sense:

Definition 7.18 *A seminorm* $\|\ \|$ *on a vector space* $\mathcal{F}$ *of functions is called* **solid** *if* $|f| \le |g|$ *implies* $\|f\| \le \|g\|$.

Making this definition is not meant to intimate that most seminorms are solid; in integration theory, though, the naturally occurring seminorms are, and their solidity plays a fundamental rôle.

Supplements and Additional Exercises

Exercise 7.19 Let $(\mathcal{F}, \|\ \|)$ be a seminormed vector space and $F : \mathcal{F} \to \mathbb{R}$ a function. Define the notions of continuity of F at $f \in \mathcal{F}$ and of uniform continuity of F. Show that the uniform limit of a sequence of uniformly continuous functions is uniformly continuous.

Exercise 7.20 Prove that the Fourier transform is one–to–one. (Recall that the Fourier transform $\mathcal{F}f$ of a continuous Riemann integrable function $f : \mathbb{R} \to \mathbb{R}$ is defined by

$$\mathcal{F}f(\alpha) = \int_{-\infty}^{\infty} f(x) e^{i\alpha x}\, dx, \ \alpha \in \mathbb{R}\,.)$$

The following project tests the comprehension of this chapter by way of a generalization that was originally carried out by Jordan.

Project 7.21 *Let* S *be any set,* $\mathcal{E}$ *a ring and vector lattice closed under chopping of bounded functions on* S*, and* $\int : \phi \mapsto \int \phi \in \mathbb{R}$ *an increasing linear map on* $\mathcal{E}$*. Such a pair* $(\mathcal{E}, \int)$ *is called a positive elementary integral on* S*. Assume also that* $\mathcal{E}$ *is self–confined* *(see page 8)**.*
(i) Develop the Riemann integral of $(\mathcal{E}, \int)$ *and establish its permanence properties.*
(ii) As an example treat the Riemann integral on $\mathbb{R}^2$ *and on* $\mathbb{R}^n$*.*

Chapter II

Extension of the Integral

Historical Note. Riemann introduced his integral in the context of *Fourier series.* It happens in numerous contexts, from engineering to number theory, that one wishes to write a function f defined on the interval $(-\pi, \pi]$ as a superposition of simple trigonometric functions:

$$f(x) = \frac{a_0}{2} + \sum_{n=1}^{\infty} a_n \cos(nx) + b_n \sin(nx) \;, \tag{1}$$

i.e. as a *Fourier series.* The reader might well have come across them in advanced Calculus or in a course on differential equations. The problem is to determine which functions can be represented as in Equation (1); in which sense they are so represented, that is to say, in which sense the partial sums, the *trigonometric polynomials*

$$f_N(x) = \frac{a_0}{2} + \sum_{\nu=0}^{N} a_\nu \cos(\nu x) + b_\nu \sin(\nu x)$$

converge to f as $N \to \infty$; and then to find the coefficients a_n, b_n. Assume for argument's sake that the f_N converge, say pointwise, to f. The coefficients should then be given by

$$a_n = \frac{1}{\pi} \int_{-\pi}^{\pi} f(x) \cos(nx)\, dx \quad \text{and} \quad b_n = \frac{1}{\pi} \int_{-\pi}^{\pi} f(x) \sin(nx)\, dx \;. \tag{2}$$

The reason is this: for $N \geq n$

$$\int_{-\pi}^{\pi} f_N(x) \cdot \cos(nx)\, dx = \pi a_n \quad \text{and} \quad \int_{-\pi}^{\pi} f_N(x) \cdot \sin(nx)\, dx = \pi b_n \;. \tag{3}$$

This follows from the well–known orthogonality relations

$$\int_{-\pi}^{\pi} \cos(\nu x) \cos(nx)\, dx = \int_{-\pi}^{\pi} \sin(\nu x) \sin(nx)\, dx = \begin{cases} 0 & \text{if } \nu \neq n \\ \pi & \text{if } \nu = n \end{cases}$$

and

$$\int_{-\pi}^{\pi} \cos(\nu x) \sin(nx)\, dx = 0 \;.$$

K. Bichteler, *Integration – A Functional Approach*, Modern Birkhäuser Classics,
DOI 10.1007/978-3-0348-0055-6_2, © Birkhäuser Verlag 1998

If we let $N \to \infty$ in Equation (3), the integrals tend to

$$\int_{-\pi}^{\pi} f(x)\cos(nx)\,dx \quad \text{and} \quad \int_{-\pi}^{\pi} f(x)\sin(nx)\,dx \;,$$

respectively. Voilà.

This argument uses a theorem about the interchange of limit and integral that does not exist! What we need here is a theorem like this: "If (f_n) is a sequence of integrable functions that converges pointwise to f then f is integrable and $\int f_n \xrightarrow[n\to\infty]{} \int f$." As mathematicians in the late 19th century were looking for results of this nature, it became soon clear that any theorems one might conjecture and that would really help are simply not true (see Exercise I.6.6). Lebesgue cut the Gordian knot in 1902 by proposing a notion of integral that encompasses Riemann's and features limit theorems strong enough to advance the theory of Fourier series considerably. For example, the limit Theorem I.6.5 stays, even when uniform convergence in (a) is replaced by mere pointwise convergence — this is Lebesgue's celebrated Dominated Convergence Theorem (DCT). (The theorem quoted at the beginning of this paragraph is not true in his theory, though, without the domination condition (b); there exists no notion of integral for which it would be.) The DCT produces vast numbers of new integrable functions, namely, the pointwise limit of any dominated sequence of functions that have been shown to be integrable. In fact, good limit theorems go hand–in–hand with abundance of integrable functions.

Our strategy — or rather Daniell's strategy — towards an integral extension that provides more integrable functions is to replace Jordan's mean $\| \ \|^\natural$ with a smaller semi–norm $\| \ \|^*$, Daniell's mean, that still majorizes the elementary integral; this makes it easier for a function f to be approximable by step functions: f may be such that $\|f-\phi\|^\natural > 1$ for all $\phi \in \mathcal{E}$, yet there may be a sequence (ϕ_n) in $\mathcal{E}$ with $\|f-\phi_n\|^* \to 0$; f is then not Riemann integrable, yet is integrable in the new sense.

Daniell's construction of $\| \ \|^*$ is quite similar to that of $\| \ \|^\natural$. The Riemann upper integral $\int^\natural$ is simply replaced by a better suited one, the Daniell upper integral $\int^*$. The Daniell mean $\| \ \|^*$ is then defined by $\|f\|^* = \int^*|f|$, just as $\| \ \|^\natural$ was defined by $\|f\|^\natural = \int^\natural|f|$, and from there we proceed as in the previous chapter. Not only will $\| \ \|^*$ be finite on some functions that are unbounded or have unbounded support — thus subsuming, for instance, positive improperly Riemann integrable functions into the theory —, functions much wilder than Riemann integrable ones can be approximated by step functions in the sense of $\| \ \|^*$. In fact, the closure of the step functions under Daniell's mean $\| \ \|^*$, the new class of integrable functions, is so rich that it is quite hard to exhibit a function not in it (Section 8).

Daniell's seminorm $\| \ \|^*$ is again solid and majorizes the elementary integral. It has an additional continuity property that Jordan's seminorm $\| \ \|^\natural$ does not share: for every sequence (f_n) of positive functions,

$$\Big\| \sum_{n=1}^{\infty} f_n \Big\|^* \leq \sum_{n=1}^{\infty} \|f_n\|^* . \tag{CSA}$$

This is reasonably called ***countable subadditivity***. It is the countable subadditivity of $\| \ \|^*$ in conjunction with the other properties that leads to the beautiful limit theorems of Lebesgue's integral.

It turns out that countably subadditive solid function seminorms arise in other contexts, not necessarily from Daniell upper integrals, so that the limit theorems are available for them, as well. We shall make it a point to state in every section which properties of $\| \ \|^*$ are actually used in the proofs. This will allow us later to invoke the results for means such as the p–norms, without having to prove them again.

II.1 Σ–additivity

Before revamping Riemann's theory of integration with all its intuitive appeal one would try, of course, to prove limit theorems better than the Dominated Uniform Convergence Theorem I.6.5. One might imagine that the next result was discovered this way. It is an extremely modest result, applying as it does only to a sequence of step functions whose pointwise and monotone limit is *a priori* known to be a step function again. Yet the whole development of the Lebesgue integral rests on it.

Lemma 1.1 *(i) Let (ϕ_n) be an increasing sequence of step functions on the line whose pointwise supremum is again a step function ϕ. Then*

$$\lim_n \int \phi_n = \int \lim_n \phi_n = \int \phi .$$

(ii) Let (δ_n) be a sequence of positive step functions on the line whose pointwise sum exists and is a step function ϕ again. Then

$$\sum_n \int \delta_n = \int \sum_n \delta_n = \int \phi .$$

(iii) Let (ψ_n) be a decreasing sequence of step functions with pointwise infimum zero. Then

$$\int \psi_n \xrightarrow[n\to\infty]{} 0 .$$

Proof. These three statements are in a trivial way consequences of each other: if (iii) is true then we apply it to the sequence $(\phi - \phi_n)$ and obtain (i); (i) applied to the sequence of partial sums $\sum_{\nu=1}^{n} \delta_\nu$ yields (ii); and (ii) applied to the decreasing sequence $(\psi_n - \psi_{n+1})$ says that

$$\int \psi_1 = \sum_{\nu=1}^{\infty} \int (\psi_\nu - \psi_{\nu+1}) = \lim_n \sum_{\nu=1}^{n} \int (\psi_\nu - \psi_{\nu+1}) = \int \psi_1 - \lim_n \int \psi_{n+1} \, ,$$

and implies $\int \psi_n \xrightarrow[n\to\infty]{} 0$.

So we prove (iii). Let $\epsilon > 0$ be given and let $(-M, M]$ be an interval containing the carrier $[\psi_1 > 0]$ of ψ_1. All the ψ_n vanish off this interval. Let $x_n^1, \ldots, x_n^{I(n)}$ be the points where ψ_n jumps, $n = 1, 2, \cdots$, and consider the open sets

$$B_n = \bigcup_{m=0}^{n} \bigcup_{i=1}^{I(m)} (x_m^i - \epsilon 2^{-m-i-2}, x_m^i + \epsilon 2^{-m-i-2}) \, .$$

They clearly increase with n. Next consider the sets

$$U_n = B_n \cup [\psi_n < \epsilon] \, .$$

These sets are again open. For if $x \in B_n$ then B_n itself is a neighborhood of x contained in U_n; and if $x \in U_n \backslash B_n$ then ψ_n does not jump at x, so that $\psi_n < \epsilon$ in a whole neighborhood of x. The sets U_n clearly increase with n. Their union is the whole line. For if $x \in \mathbb{R}$ then $x \in [\psi_n < \epsilon] \subset U_n$ provided the index n is so high that $\psi_n(x) < \epsilon$. Such indices exist since $\psi_n(x) \downarrow_{n\to\infty} 0$. There will exist an index N such that U_N contains the *compact* set $[-M, M]$. Thus for all $n \geq N$

$$[-M, M] \subset [\psi_n < \epsilon] \cup \widetilde{B}_n \, , \tag{1.1}$$

where
$$\widetilde{B}_n = \bigcup_{m=0}^{n} \bigcup_{i=1}^{I(m)} (x_m^i - \epsilon 2^{-m-i-2}, x_m^i + \epsilon 2^{-m-i-2}]$$

is a set of $\mathcal{E}[\mathbb{R}]$ slightly bigger than B_n. Equation (1.1) says that $\psi_n < \epsilon$ off $\widetilde{B}_n$ for $n \geq N$. Now $\widetilde{B}_n$ is a set of measure less than

$$\sum_{n=0}^{\infty} \sum_{i=1}^{\infty} 2^{-n-i-1} \epsilon = \epsilon \, .$$

The set $G_n = (-M, M] \backslash \widetilde{B}_n$ belongs to $\mathcal{E}[\mathbb{R}]$, and $\psi_n = G_n \cdot \psi_n + \widetilde{B}_n \cdot \psi_n$. Thus

$$\begin{aligned} \int \psi_n(x)\, dx &= \int_{G_n} \psi_n(x)\, dx + \int_{\widetilde{B}_n} \psi_n(x)\, dx \\ &\leq \epsilon \cdot \lambda(G_n) + \epsilon \cdot \sup_x \psi_n(x) \leq \epsilon \cdot \Big(M + \|\psi_1\|_u\Big) \end{aligned}$$

for $n \geq N$. Since $\epsilon > 0$ was arbitrary, (iii) follows. ∎

Statement (i) is commonly paraphrased as "*the elementary Riemann integral is σ–continuous*," and statement (ii) as "*the elementary Riemann integral is σ–additive*." (The prefix σ– connotes countable behavior in increasing circumstances.) Statement (iii) can be read as saying "*the elementary Riemann integral is δ–continuous at zero*." (The prefix δ– connotes countable behavior in decreasing circumstances.)

Exercise 1.2 Properties (i) – (iii) are equivalent with σ–continuity at zero.

Lemma 1.1 is a rather modest limit result. Yet it is fair to say that Lebesgue's theory amounts to parlaying it into most satisfying limit theorems for his new extension of the integral.

II.2 Elementary Integrals

It was soon noticed that Lebesgue's methods cover not only the integral on the line, but also the integral on the plane or on $\mathbb{R}^n$. They make use only of the fact that the step functions $\mathcal{E}[\mathbb{R}]$ form a lattice ring and that the "elementary integral" $\int : \mathcal{E}[\mathbb{R}] \to \mathbb{R}$ is linear, positive, and σ–additive. They do, in particular, not rely on the nature of the "elementary functions" in $\mathcal{E}[\mathbb{R}]$. In other words, they apply to the general positive σ–additive elementary integral:

Definition 2.1 *(i) An* ***elementary integral*** *is a pair $(\mathcal{E}, m)$ consisting of a lattice ring $\mathcal{E}$ of bounded functions on some set, and of a linear map*

$$m : \mathcal{E} \to \mathbb{R} .$$

The functions in $\mathcal{E}$ are the ***elementary integrands*** *or* ***elementary functions****. If we need to refer to the set on which they live, we call it the* ***ambient set****. The linear map m by itself is — in slight abuse of the language — again called the* ***elementary integral****. It is sometimes convenient to write $\int \phi \, dm$ or $\int \phi(x)\, m(dx)$ for $m(\phi)$, or even simply $\int \phi$ if there is no doubt which linear map $m : \mathcal{E} \to \mathbb{R}$ is meant:*

$$m(\phi) = \int \phi \, dm = \int \phi(x)\, m(dx) = \int \phi , \qquad \phi \in \mathcal{E} .$$

The letter λ is reserved for the Lebesgue integral. Whenever we want to stress that we are in the particular context of the usual integral on the line we write $(\mathcal{E}[\mathbb{R}], \lambda)$, $(\mathcal{E}[\mathbb{R}], \int dx)$, $(\mathcal{E}[\mathbb{R}], \int d\lambda)$, $(\mathcal{E}[\mathbb{R}], \int \lambda(dx))$, and talk about the ***elementary Lebesgue integral****.*
(ii) The elementary integral $(\mathcal{E}, m) = (\mathcal{E}, \int)$ is ***positive*** *if $0 \le \phi \in \mathcal{E}$ implies that $0 \le m(\phi)$ or, equivalently, if $\int \phi \le \int \psi$ for $\phi \le \psi$ in $\mathcal{E}$.*

(iii) The elementary integral $(\mathcal{E}, \int)$ *is* $\boldsymbol{\sigma}$***–additive*** *if*

$$\int \sum \delta_n = \sum \int \delta_n$$

for every sequence (δ_n) *of positive elementary integrands whose pointwise sum exists and is an elementary integrand — the proof of Lemma 1.1 shows that this is equivalent with* δ*–continuity at zero, and also with* σ*–continuity: for every increasing sequence* (ϕ_n) *of elementary integrands whose pointwise supremum is again an elementary integrand*

$$\lim_n \int \phi_n = \int \lim_n \phi_n = \int \phi \,. \tag{2.1}$$

Exercise 2.2* The σ–additivity, σ–continuity, and δ–continuity of an elementary integral are equivalent, whether it is positive or not.

Exercise 2.3 The elementary Riemann integral on the plane (Exercise I.4.8) is σ–additive. So is the elementary Riemann integral on $\mathbb{R}^n$.

Lemma 1.1 states that the usual integral $(\mathcal{E}[\mathbb{R}], \int \, dx)$ of step functions on the line is a positive σ–additive elementary integral. As pointed out above, Lebesgue's or Daniell's extension theories of the elementary Lebesgue integral on the line use only this fact. The Lebesgue integral does have special properties having to do with the structure of the ambient space $\mathbb{R}$ as a topological field, but these do not enter the extension theory. The latter applies to every positive σ–additive elementary integral. We shall therefore develop right away the integration theory of the general positive σ–additive elementary integral. Apart from the obvious gain in generality, the arguments become clearer this way.

The reader interested only in the Lebesgue integral on the line can skip the remainder of this section, provided he understands $\mathcal{E}$ to mean $\mathcal{E}[\mathbb{R}]$ and $m(\phi) = \int \phi$ to mean $\int \phi(x)\, dx$ in the sequel. For the reader interested in the full generality it may still be helpful to visualize the usual integral $(\mathcal{E}[\mathbb{R}], \int \, dx)$ on the line during the arguments, or better yet the integral $(\mathcal{E}[\mathbb{R}^2], \int \, dxdy)$ in the plane — especially if he has done Exercise 2.3.

Mathematicians, too, are subject to temptations, and one of the more insidious ones is to generalize for generalization's sake — thus producing what in the community is called "generalized abstract nonsense." We owe it to the reader to dispel his suspicion that Definition 2.1 does just that. To this end we exhibit in the remainder of this section two rich classes of examples of positive σ–additive elementary integrals, the Radon measures and the (linear extensions of) set functions. Both occur abundantly in nature.

Radon Measures

Let S be a locally compact Hausdorff space[1]. $C_{00}[S]$ denotes the collection of continuous functions on S that have compact support. This is plainly a lattice ring of bounded functions and thus qualifies for the domain $\mathcal{E}$ of an elementary integral. A positive linear functional $m : C_{00}[S] \to \mathbb{R}$ is called a ***positive Radon measure*** — they abound, as we shall see later (Example 2.21 and Exercise IV.4.8). In keeping with the notation of Definition 2.1 we also write $m(\phi) = \int \phi \, dm$. Radon measures are the prime tools in the analysis of the ubiquitous spaces $C[K]$, K compact.

Lemma 2.4 *A positive Radon measure m is σ–additive.*

Proof. Let (ϕ_n) be an increasing sequence in $C_{00+}[S]$ with pointwise supremum $\phi \in C_{00}[S]$. By Dini's Lemma I.3.1, the sequence (ϕ_n) converges to ϕ uniformly on the support of ϕ. There exists a positive function $\psi \in C_{00}[S]$ that equals 1 on the support of ϕ [2] — in the language of Proposition I.4.3, ϕ is confined by ψ. From here on we simply repeat the proof of the Dominated Uniform Convergence Theorem I.6.5: the order relation $|\phi - \phi_n| \le \psi \cdot \|\phi - \phi_n\|_u$ in conjunction with the positivity of m implies

$$|m(\phi) - m(\phi_n)| = m(\phi - \phi_n) \le m(\psi) \cdot \|\phi - \phi_n\|_u \xrightarrow[n\to\infty]{} 0 . \qquad \blacksquare$$

Exercise 2.5 Let S be an open or closed subset of $\mathbb{R}^n$ and $K \subset S$ compact. There exists a continuous function $\psi : S \to [0,1]$ of compact support which equals 1 on K.

Exercise 2.6 The Riemann integral restricted to $C_{00}[\mathbb{R}^n]$ is a positive Radon measure.

Exercise 2.7* The proof of the σ–additivity in Lemma 2.4 actually yields a little more: Let Φ be an increasingly directed subfamily of C_{00+}; that is to say, given any two members of Φ there is a third one that exceeds both. Assume the pointwise supremum of Φ is a function $\psi \in C_{00}[S]$. Then $m(\sup \Phi) = \sup_{\phi\in\Phi} m(\phi) = \lim_{\phi\in\Phi} m(\phi)$.

The behavior of Radon measures shown in Exercise 2.7 gives rise to the following

Definition 2.8 (Order–continuous elementary integrals) *An elementary integral $(\mathcal{E}, \int)$ is **order–continuous** if for every increasingly directed subset Φ of elementary integrands whose pointwise supremum $\sup \Phi$ is also an elementary integrand*

$$m(\sup \Phi) = \lim_{\phi\in\Phi} m(\phi) .$$

---

[1] The reader not familiar with general topology may take this to mean an open or a closed subset of $\mathbb{R}^n$. In such a set any two distinct points clearly have disjoint compact neighborhoods — this accounts for the words "Hausdorff" and "locally compact."

[2] Urysohn's lemma provides it — see any textbook on point set topology. The reader not versed in point set topology should do Exercise 2.5.

Order–continuous elementary integrals have a slightly superior integration theory (see, however, Exercise III.5.11 on p. 90). It is left to the reader to establish this in the exercises.

Set Functions

The next class of examples, the positive σ–additive set functions, occur naturally on $\mathbb{R}^n$ and in elementary models of probability theory.

A non–void collection $\mathcal{R}$ of subsets of some set S is a ***ring of sets*** if it is closed under taking finite unions and relative complements — and then under taking finite intersections: $A \cap B = A \setminus (A \setminus B)$. If the ring $\mathcal{R}$ contains the ambient set S then it is called an ***algebra of sets***.

A ***measure*** on $\mathcal{R}$ is a map $\mu : \mathcal{R} \to \mathbb{R}$ that is additive: for any two disjoint sets $A, B \in \mathcal{R}$,

$$\mu(A \cup B) = \mu(A) + \mu(B) .$$

μ is ***positive*** if $\mu(A) \geq 0$ for every $A \in \mathcal{R}$ and ***σ–additive*** if

$$\mu\Big(\bigcup A_n\Big) = \sum \mu(A_n)$$

for any countable collection $\{A_n\}$ of mutually disjoint sets in $\mathcal{R}$ whose union belongs to $\mathcal{R}$. A simple argument as in Lemma 1.1 or Exercise 2.2 lets us rephrase the σ–additivity as

σ–continuity: $\quad \mathcal{R} \ni A_n \uparrow A \in \mathcal{R} \implies \mu(A_n) \to \mu(A)$

or as δ–continuity at $\emptyset$: $\quad \mathcal{R} \ni A_n \downarrow \emptyset \implies \mu(A_n) \to 0 .$

Exercise 2.9* The two previous conditions are reasonably called σ–continuity and δ–continuity at zero, respectively. Show that σ–additivity, σ–continuity, and δ–continuity of a measure are equivalent, whether it is positive or not.

Exercise 2.10 A ring contains the empty set, and the empty set has measure zero.

Exercise 2.11 A non–void collection of subsets closed under taking finite unions is a ring if it is also closed under taking relative complements and an algebra if it is also closed under taking complements.

Exercise 2.12 Let S be a set. (i) The intersection of any collection of rings of subsets of S is a ring of subsets. (ii) Let $\mathcal{C}$ be a non–void family of subsets of S. The collection of all rings containing $\mathcal{C}$ is not void. The intersection of this collection is called the ring generated by $\mathcal{C}$. Let us denote it by $[\mathcal{C}]$. (iii) If $\mathcal{C}$ is finite then so is $[\mathcal{C}]$, and if $\mathcal{C}$ is countable then so is $[\mathcal{C}]$.

A measure $(\mathcal{R}, \mu)$ gives rise to an elementary integral by the simple expedient of extension by linearity: A *step function over* $\mathcal{R}$ is, of course, a function ϕ that takes only finitely many values, each non–zero value on a set from $\mathcal{R}$. That is to say, the "level sets" $[\phi = r]$ belong to $\mathcal{R}$ for every $r \neq 0$, all but finitely many of them being void. Let us denote by $\mathcal{E}[\mathcal{R}]$ the collection of step functions over $\mathcal{R}$. There is an immediate and intuitive way to define the μ–integral of a step function, namely by the time–honored formula $\sum$ *height–of–step* $\times$ *μ–size–of–step*:

$$\int \phi \, d\mu = \sum_{r \in \mathbb{R}} r \cdot \mu([\phi = r]) \ . \tag{2.2}$$

This is a finite sum. It is not quite obvious that $\mathcal{E}[\mathcal{R}]$ is, indeed, a lattice ring and that the integral is linear.

Proposition 2.13 *Let μ be a measure on the ring $\mathcal{R}$ of sets.*
(i) The step functions $\mathcal{E}[\mathcal{R}]$ are exactly the linear span of the (indicator functions of the) sets in $\mathcal{R}$ and form a lattice ring.
(ii) The map $\int d\mu$ of Equation (2.2) is the unique linear extension of μ to $\mathcal{E}[\mathcal{R}]$. That is to say, $\int d\mu : \mathcal{E}[\mathcal{R}] \to \mathbb{R}$ is linear, agrees on $\mathcal{R} \subset \mathcal{E}[\mathcal{R}]$ with μ, and is the only map with these two properties. For this reason we also write $\mu(\phi)$ for $\int \phi \, d\mu$ when $\phi \in \mathcal{E}[\mathcal{R}]$.

For the proof a little lemma is needed:

Lemma 2.14 *For every finite collection $\mathcal{I} = \{A_1, \ldots, A_I\}$ of sets in $\mathcal{R}$ there is another finite collection $\mathcal{C} = \{C_1, \ldots, C_N\}$ of* **mutually disjoint** *sets in $\mathcal{R}$ such that every $A_i \in \mathcal{I}$ is the union of sets in $\mathcal{C}$:*

$$A_i = \bigcup \{C_n : \ C_n \subset A_i\} \quad \textit{for} \ \ 1 \leq i \leq I \ . \tag{2.3}$$

Proof. This is obvious if $I = 1$. If it is true for $I - 1$, and if $\mathcal{C}' = \{C_1, \ldots, C_{N'}\}$ is the finite collection of mutually disjoint sets in $\mathcal{R}$ such that (2.3) is satisfied for $1 \leq i \leq I - 1$, then

$$\mathcal{C} = \{C_n \cap A_I : 1 \leq n \leq N'\} \cup \{C_n \backslash A_I : 1 \leq n \leq N'\} \cup \Big\{ A_I \setminus \bigcup_{n=1}^{N'} C_n \Big\}$$

is a collection of at most $2N' + 1$ mutually disjoint sets from $\mathcal{R}$ satisfying (2.3).

Now to the proof of Proposition 2.13. To show that the step functions over $\mathcal{R}$ form a lattice ring let ϕ_1, ϕ_2 be two of them. Let $\mathcal{I}$ be the collection of non–void level sets of ϕ_1 and ϕ_2 and $\mathcal{C}$ the disjoint collection provided by Lemma 2.14. Note that we can write[3]

$$\begin{aligned} \phi_i &= \sum_{r \in \mathbb{R}} r \cdot [\phi_i = r] = \sum_{r \in \mathbb{R}} r \cdot \sum \{C \in \mathcal{C} : \ C \subset [\phi_i = r]\} \\ &= \sum_{r, C \subset [\phi_i = r]} r \cdot C = \sum_{r, C \subset [\phi_i = r]} \phi_i(C) \cdot C = \sum_{C \in \mathcal{C}} \phi_i(C) \cdot C \end{aligned}$$

---

[3] ϕ_i is clearly constant on each $C \in \mathcal{C}$, and $\phi_i(C)$ denotes the value it takes there.

for $i=1,2$. That is to say, ϕ_i is a linear combination of (the indicator functions of) the sets in $\mathcal{C}$. Since the latter are disjoint, it is evident that the sum, infimum, etc. of ϕ_1 and ϕ_2 are step functions over $\mathcal{C}$: $\mathcal{E}[\mathcal{R}]$ is a lattice ring. There is a simple expression for the integral in terms of $\mathcal{C}$:

$$\int \phi_i = \sum_r r \cdot \mu([\phi_i = r]) = \sum_r r \cdot \sum_{C \subset [\phi_i = r]} \mu(C) = \sum_{C \in \mathcal{C}} \phi_i(C) \cdot \mu(C) \,.$$

The linearity of the integral is obvious from this. ∎

Exercise 2.15* The measures on a fixed ring of sets form a vector space under pointwise operations. So do the elementary integrals on the step functions over that ring. The map that associates with a measure on the ring its extension on the step functions is a linear isomorphism that preserves the order: $\mu \ge 0 \iff \int d\mu \ge 0$.

The extension by linearity of a measure on a ring to an elementary integral on the step functions over that ring is such a simple procedure that we shall perform it automatically and without mention whenever we encounter a measure μ. We take the liberty to denote the extension again by μ if that is notationally convenient:

$$\mu : \mathcal{R} \xrightarrow{\text{additive}} \mathbb{R} \qquad \text{gives rise to} \qquad \mu : \mathcal{E}[\mathcal{R}] \xrightarrow{\text{linear}} \mathbb{R} \,.$$

The extension preserves σ–additivity, at least for positive measures:

Proposition 2.16 *Let μ be a positive measure on a ring $\mathcal{R}$ of sets. Then its linear extension $\int d\mu$ is σ–additive if and only if μ is.*

Proof. The necessity is obvious. We shall prove the suffiency in the form of Equation (2.1): if (ϕ_n) is a sequence in $\mathcal{E}[\mathcal{R}]$ with pointwise supremum $\phi \in \mathcal{E}[\mathcal{R}]$ then $\int \phi_n \, d\mu \xrightarrow[n\to\infty]{} \int \phi \, d\mu$. This is evident from the inequality

$$\begin{aligned}\int (\phi - \phi_n) \, d\mu &= \int (\phi - \phi_n) \cdot [\phi - \phi_n \le \epsilon] \, d\mu + \int (\phi - \phi_n) \cdot [\phi - \phi_n > \epsilon] \, d\mu \\ &\le \mu([\phi > 0]) \cdot \epsilon + \|\phi\|_u \cdot \mu([\phi - \phi_n > \epsilon]) \,.\end{aligned}$$

This number can be made arbitrarily small by the choice first of $\epsilon > 0$ and then of n. Indeed, the sets $[\phi - \phi_n > \epsilon] \in \mathcal{R}$ decrease to the void set, which has measure zero, and their μ–measure thus tends to zero. ∎

Exercise 2.17* Why did we stipulate that the elementary sets that can be measured should form a ring, and that the measure be additive on it? The following is meant to shed some light on this question. Let $\mathcal{C}$ be a collection of subsets of some set S. A step function or simple function over $\mathcal{C}$ is, naturally, a function f that takes only finitely many values, all of them finite, and such that the sets $[f = r]$ belong to $\mathcal{C}$ for $0 \ne r \in \mathbb{R}$.

Show that the step functions over $\mathcal{C}$ form a vector space precisely if $\mathcal{C}$ is a ring; and that they then form a lattice ring. Show that this ring of functions contains the constant 1, i.e. is an algebra of functions, if and only if $\mathcal{C}$ is an algebra of sets.
A map $\mu : \mathcal{C} \to \mathbb{R}$ has a linear extension to the step functions over $\mathcal{C}$ if and only if it is additive on $\mathcal{C}$.

Example 2.18 The usual length function λ on $\mathcal{A}[\mathbb{R}]$ is an instance of a positive σ–additive measure on a ring, by Lemma 1.1; so is the area on the ring $\mathcal{A}[\mathbb{R}^2]$ of sets in $\mathcal{E}[\mathbb{R}^2]$ of Exercise I.4.8 and the volume of boxes and their finite unions in $\mathbb{R}^n$.

Exercise 2.19* (Distribution Functions) Let F be an increasing function on the line. For any $a \le b$ set $\mu(\,(a,b]\,) = F(b) - F(a)$. This "interval function" has a unique extension to a positive measure $\mu = dF$ on all of $\mathcal{A}[\mathbb{R}]$. μ is σ–additive if and only if F is right–continuous. The function F is called ***a distribution function*** of μ. The corresponding elementary integral is denoted by $\int \phi(x)\,F(dx)$ or $\int \phi(x)\,dF(x)$.

Example 2.20 (Probability) Here is another instance in which measures on rings of sets arise naturally. Many elementary probabilistic models for a physical system start out with modelling the states of the system by the points of a set Ω, stipulating a correspondence of observable events with an algebra $\mathcal{A}$ of subsets of Ω, and assuming that the probability of an event is modelled by the value $\mathbb{P}(A)$ of a positive measure $\mathbb{P}$ on the subset $A \subset \Omega$ corresponding to the event. The step functions ϕ over $\mathcal{A}$ are called observables and are identified with measurements on the system, and the elementary integral $\int \phi\; d\mathbb{P}$ can be defined on them as in Proposition 2.13 and is called the *expectation* and written as $\mathbb{E}[\phi]$. The problem arises to extend $\mathbb{E}$ to sufficiently many functions of the state (more *observables*) to obtain a rich calculus. To this end one requires that $\mathbb{P}$ be σ–additive—a requirement hard to justify except by its results—and then applies the integration theory below.

Example 2.21 (Probability) There is another way to make a probabilistc model for a physical system, one that comes with the σ–additivity built in. For concreteness' sake assume the system is an ear of corn. There are many measurements one can perform: length, diameter, number of kernels, weight, length plus weight, diameter plus 3 times the number of kernels, etc. These measurements form in an obvious way an "abstract" algebra $\mathcal{M}$. For every $m \in \mathcal{M}$ let $\|m\|$ denote the supremum of all values that the measurement m could ever take, over all possible ears of corn. Clearly $\|\;\|$ is a norm on $\mathcal{M}$ with $\|m_1 \cdot m_2\| \le \|m_1\| \cdot \|m_1\|$ and $\|m^2\| = \|m\|^2$. A commutative normed algebra $(\mathcal{M}, \|\;\|)$ having this property is called a real C^*–algebra, and there is a theorem that it can be realized as the space of continuous functions on a compact space, with $\|\;\|$ corresponding to the uniform norm $\|\;\|_u$. In other words, there is a compact space K and an isometric isomorphism of $\mathcal{M}$ with (a dense subalgebra of) $C[K]$. Now the expected value of a measurement[4] m evidently should be increasing and depend linearly on $m \in \mathcal{M}$. Transported to $C[K]$, it is modelled by a positive Radon measure $\phi \mapsto \mathbb{E}[\phi]$ on K. As such it is automatically σ–additive.

Example 2.22 Here is an example of an elementary integral that can be viewed both as a Radon measure and as coming from a set function. Let S be any set, let $\mathcal{R}$ denote the collection of all finite subsets, and $\mathcal{E}$ the collection of functions that vanish at all but finitely many points. Clearly, $\mathcal{E}$ are the step functions over $\mathcal{R}$, and $\mathcal{R}$ are the idempotent functions of $\mathcal{E}$. For $A \in \mathcal{R}$ let $\mu(A)$ be the number of points in A. This is plainly a σ–

[4] In the so–called frequency model this would be the limit of the average of the measurement m taken over ever bigger samples of ears of corn.

additive measure. Its linear extension to $\mathcal{E}$ is given by the formula

$$\int \phi \, d\mu = \sum_{s \in S} \phi(s) .$$

If S is given the discrete topology by declaring every set to be open, then $\mathcal{E} = C_{00}[S]$ and $\int \, d\mu$ is a Radon measure. In any case, μ is known as ***counting measure***.

Remark 2.23 In the examples where we established σ–additivity (Lemma 1.1, Exercise I.4.8, Lemma 2.4, Exercise 2.19, and Example 2.21), it was, with mind–numbing monotonicity, due to compactness in the ambient space S. There are, in fact no natural instances of σ–additivity that are not ultimately due to compactness. This has led Bourbaki [2] to identify integration theory with the theory of Radon measures. This view does not have a large following, because keeping track of the compact sets in the ambient space is often too cumbersome, in particular in the case of probability. The preponderant view today is to establish σ–additivity using compactness, rejoice in it, and to invoke the topology thereafter only when it is inevitable.

II.3 The Daniell Mean

In the remainder of the chapter, $(\mathcal{E}, \int)$ is a fixed positive σ–additive elementary integral [5]. If there is need to mention the set on which the functions of $\mathcal{E}$ live we refer to it as the ***ambient set***.

The Construction of the Daniell Upper Integral $\int^*$, replacement for the Riemann upper integral $\int^\natural$, is in two steps. In the first step it is defined only for functions h that are pointwise suprema of countable families in $\mathcal{E}$. Let $\mathcal{E}^\uparrow$ denote their collection: a function h belongs to $\mathcal{E}^\uparrow$ if there exists a sequence (ϕ_n) of elementary functions whose pointwise supremum equals h. Replacing ϕ_n by $\phi_1 \vee \ldots \vee \phi_n$ we see that such a sequence can be chosen to be increasing. It is understood that a function in $\mathcal{E}^\uparrow$ may take the value ∞. The upper integral of a function h in $\mathcal{E}^\uparrow$ is defined by

$$\int^* h = \sup\{ \int \phi : h \geq \phi \in \mathcal{E} \} . \tag{3.1}$$

If the set of numbers in the brackets is unbounded this is to mean the symbol $+\infty$. For an arbitrary function f set

$$\int^* f = \inf\{ \int^* h : f \leq h \in \mathcal{E}^\uparrow \} . \tag{3.2}$$

---

[5] As pointed out at the beginning of Section 2, the reader may take this to be the usual integral $(\mathcal{E}[\mathbb{R}], \int \, dx)$ on the line.

If the set of numbers in the brackets is void, i.e., if f is not majorized by any function in $\mathcal{E}^{\uparrow}$, this is to mean the symbol $+\infty$. We refer to this up–and–down way to define the Daniell upper integral as "*Daniell's up–and–down procedure.*"

The reader might be wondering why we should want to make these definitions or how anybody came to dream them up if they really should turn out useful. A little patience; we shall address these questions on page 45. It will facilitate the discussion to know the properties of $\int^*$.

We start with the behavior of $\int^*$ on $\mathcal{E}^{\uparrow}$.

Lemma 3.1 *(i) $\mathcal{E}^{\uparrow}$ is closed under addition, multiplication with positive scalars, and under taking finite infima and countable suprema.*
(ii) $\int^*$ is continuous along increasing sequences of $\mathcal{E}^{\uparrow}$; that is to say, for any increasing sequence (h_n) in $\mathcal{E}^{\uparrow}$ with pointwise supremum h — which by (i) belongs to $\mathcal{E}^{\uparrow}$ —

$$\int^* h = \lim_{n\to\infty} \int^* h_n = \sup_n \int^* h_n \ . \tag{$*$}$$

(iii) $h \mapsto \int^* h$ is positive–homogeneous and additive on $\mathcal{E}^{\uparrow}$.

Proof. (i): Let $h_1, h_2, \ldots$ be a countable collection of functions in $\mathcal{E}^{\uparrow}$. Every one of the h_n is the supremum of a countable collection in $\mathcal{E}$:

$$h_n = \sup\{\phi_{n,k} : k = 1, 2, \ldots\} \ , \qquad \text{say.}$$

Then $h_1 + h_2$ is the supremum of $\{\phi_{1,k} + \phi_{2,k} : k \in \mathbb{N}\} \subset \mathcal{E}$, while $r{\cdot}h_1$ is the supremum of $\{r{\cdot}\phi_{1,k}\} \subset \mathcal{E}$, provided $r \in \mathbb{R}$ is positive. Next, $\sup_n h_n$ is the pointwise supremum of the countable collection $\{\phi_{n,k} : k, n = 1, 2, \ldots\} \subset \mathcal{E}$. Lastly, observe that h_n is the pointwise limit of the *increasing* sequence

$$\widehat{\phi}_{n,k} = \bigvee_{i=1}^{k} \phi_{n,i} \ \in \mathcal{E} \ , \qquad\qquad k = 1, 2, \cdots.$$

Clearly $h_1 \wedge h_2 = \sup_k \widehat{\phi}_{1,k} \wedge \widehat{\phi}_{2,k}$ belongs to $\mathcal{E}^{\uparrow}$.
(ii): Since $\int^*$ is clearly increasing, the limit in $(*)$ equals the supremum. The inequality $\int^* h \ge \sup \int^* h_n$ is obvious. Only the reverse inequality needs proving. Let $\int^* h > r \in \mathbb{R}$. There are an elementary function $\phi \le h$ with $\int \phi > r$ and countable collections $\{\phi_{n,k} : k \in \mathbb{N}\}$ of elementary functions with pointwise supremum h_n. The elementary functions

$$\widehat{\phi}_n = \phi \wedge \left(\bigvee_{1\le m,k\le n} \phi_{m,k} \right) \qquad \le h_n$$

increase pointwise to ϕ. By Lemma 1.1,

$$\lim_{n\to\infty} \int \widehat{\phi}_n > r \,.$$

As $\int^* h_n \geq \int \widehat{\phi}_n$, we have $\sup_n \int^* h_n > r$, and thus the desired inequality

$$\sup_n \int^* h_n \geq \int^* h \,.$$

(iii): The positive–homogeneity is obvious. Let then $h, h' \in \mathcal{E}^{\uparrow}$. There are sequences $(\phi_n), (\phi'_n)$ of elementary functions increasing pointwise to h, h', respectively. Clearly $(\phi_n + \phi'_n)$ increases pointwise to $h + h'$. From (ii),

$$\int^* h + h' = \lim_{n\to\infty} \int^* \phi_n + \phi'_n = \lim_{n\to\infty} \left(\int \phi_n + \int \phi'_n \right) = \int^* h + \int^* h' \,. \quad \blacksquare$$

We are in position to investigate the behavior of $\int^*$ on arbitrary functions.

Proposition 3.2 *The Daniell upper integral is increasing and positive–homogeneous. Moreover, it is countably subadditive; that is to say, for any sequence (f_n) of positive numerical functions*

$$\int^* \sum_{n=1}^{\infty} f_n \leq \sum_{n=1}^{\infty} \int^* f_n \,.$$

Lastly

$$\int^* \phi = \int \phi \quad \textit{for} \quad \phi \in \mathcal{E} \,.$$

Proof. The monotonicity and positive–homogeneity are obvious. As to the countable subadditivity, if $\sum \int^* f_n < r$ then there are functions $h_n \geq f_n$ in $\mathcal{E}^{\uparrow}$ with $\sum \int^* h_n < r$. By (ii) and (iii) of Lemma 3.1,

$$\begin{aligned} \int^* \sum_{n=1}^{\infty} f_n \leq \int^* \sum_{n=1}^{\infty} h_n &= \int^* \lim_{N\to\infty} \sum_{n=1}^{N} h_n \\ &= \lim_{N\to\infty} \int^* \sum_{n=1}^{N} h_n = \lim_{N\to\infty} \sum_{n=1}^{N} \int^* h_n < r \,. \end{aligned} \quad \blacksquare$$

Exercise 3.3 (i) Prove the monotonicity and the positive–homogeneity of $\int^*$ in all detail. (ii) Trace the countable subadditivity of $\int^*$ to the σ–additivity of $\int$. (iii) Show that $\int^{\natural}$ is not countably subadditive.

The countable subadditivity is the only one of these features of $\int^*$ that the Riemann upper integral $\int^{\natural}$ does not share. The celebrated limit theorems of the Lebesgue integral all are its consequences.

Let us take a time out to address the questions on page 43 about the provenance of the definition of $\int^*$. It takes some experience, but not much, to see that the "Little Limit Lemma" 1.1 will result in the additivity of the extension of the integral to $\mathcal{E}^{\uparrow}$, if that extension is defined by (3.1) — see Lemma 3.1 (iii). It would be too early to rejoice, because $\mathcal{E}^{\uparrow}$, though large, is not a vector space so that there is no way to so much as talk about the linearity of the extended integral on it. We have learned in Proposition I.4.5 on p. 19, though, that taking an infimum as in (I.4.3) over a functional that behaves additively results in a *subadditive* functional on *all* functions. Now, our experience with Riemann's lower integral $\int_{\natural}$ leads us to expect that replacing the "up–and–down–procedure" by a "down–and–up–procedure" will produce a *Daniell lower integral* $\int_*$ that is defined on *all* functions and is *superadditive*[6]. The set of functions on which $\int^*$ and $\int_*$ agree will be a vector space, and the common value will be both subadditive and superadditive there; i.e., it will be additive. In this way we will arrive at an extension of the integral that is defined on a large vector space of functions and is linear there. In other words, once we have realized that Lemma 1.1 will result in the additivity of the extension (3.1) on $\mathcal{E}^{\uparrow}$ we simply follow in Riemann's footsteps. This program succeeds — see Exercise 6.2 on p. 59. To summarize: some analysis of what we want to accomplish and what we have learned leads to the definitions (3.1) and (3.2) in a rather straightforward way.

Actually, we have learned in Section I.5 that it suffices to define the upper integral and to establish its subadditivity and solidity. There is no need to worry about the lower integral and its properties. A simple absolute–value sign will take care of such worries very nicely:

Definition 3.4 *As in Section I.5 we define the seminorm* $\| \ \|^*$ *associated with the upper integral* $\int^*$, *the* ***Daniell mean***, *by*

$$\|f\|^* = \int^* |f| \ \in \overline{\mathbb{R}}_+ \, .$$

We say the sequence (f_n) ***converges in*** $\| \ \|^*$ ***-mean*** *to* f *if* $\|f - f_n\|^* \to 0$.

Later we shall consider several elementary integrals $\mu = \int d\mu, \nu = \int d\nu, \ldots$ *on* $\mathcal{E}$. *The corresponding Daniell means will differ, of course, so we denote them by* $\| \ \|^{\mu}, \| \ \|^{\nu}, \ldots$. *In particular,* $\| \ \|^{\lambda}$ *denotes the Daniell mean for Lebesgue measure* λ.

We heed the lesson learned in Section I.5. Instead of using

$$-\infty < \int_* f = \int^* f < +\infty$$

[6] The reader can make sure by going through Exercise 3.14 on p. 47.

to define the integrability of f we call f integrable if it is the $\| \ \|^*$–mean limit of elementary functions. The two definitions should, of course, be equivalent, and they are (Exercise 6.2 on p. 59); the second definition is just more expeditious[7]. Before investigating integrability, though, let us list the pertinent features of $\| \ \|^*$:

Theorem 3.5 (Properties of the Daniell Mean) *(i) The number* $\|f\|^* \in [0, \infty]$ *is defined for every numerical function* f *on the ambient space. The functional* $\| \ \|^*$ *is absolute–homogeneous and solid*[8]*. Furthermore,* $\| \ \|^*$ *is countably subadditive; that is to say, for any sequence* (f_n) *of positive numerical functions*

$$\Big\| \sum_{n=1}^{\infty} f_n \Big\|^* \le \sum_{n=1}^{\infty} \|f_n\|^* . \tag{CSA}$$

(ii) The functional $\| \ \|^*$ *is finite on elementary functions. Moreover, for every sequence* (ϕ_n) *of positive elementary functions*

$$\sup_N \Big\| \sum_{n=1}^{N} \phi_n \Big\|^* < \infty \quad \textit{implies} \quad \|\phi_n\|^* \xrightarrow[n\to\infty]{} 0 . \tag{M}$$

(iii) $\| \ \|^*$ *majorizes the elementary integral: for all elementary functions* ϕ

$$\Big| \int \phi \Big| \le \|\phi\|^* .$$

Proof. (i) and (iii) are immediate from Proposition 3.2. As for (M), if $\| \sum_{n=1}^{N} \phi_n \|^* = \sum_{n=1}^{N} \int \phi_n$ is bounded independently of N then clearly $\|\phi_n\|^* = \int \phi_n \xrightarrow[n\to\infty]{} 0$. ∎

The reader might be wondering why (M) is included in the list 3.5, perfectly obvious as its proof shows it to be. The reason is that countably subadditive solid seminorms that satisfy (M) without being additive on $\mathcal{E}_+$ occur frequently. In fact, they appear with such frequency that they deserve their own name:

Definition 3.6 *Let* $\mathcal{E}$ *be a ring*[9] *of bounded functions, to be called the* ***elementary functions****, on some set. Any functional* $\| \ \|^*$ *having properties (i) and (ii) of Theorem 3.5 is called a* **mean** *for* $\mathcal{E}$.

The designation "Jordan mean" of Definition I.5.2 is really a misnomer, because $\| \ \|^\natural$ is not a mean in the sense of the definition above: it is not countably subadditive. This will not cause confusion once it has been noted.

---

[7] It has the additional advantage of applying in more general circumstances: to signed measures, which we shall meet, and to stochastic integrals and spectral measures, which the reader might meet elsewhere.

[8] Recall that this means that $\|f\|^* \le \|g\|^*$ whenever $|f| \le |g|$. Together with (ii) it implies in particular that $\||\phi|\|^* = \|\phi\|^* < \infty$ for elementary integrands ϕ.

[9] Not necessarily a vector lattice!

In the remainder of this and all of the next chapter we generally use only the fact that $\mathcal{E}$ is a ring of bounded functions and that $\| \ \|^*$ is a mean on it. In the later chapters we shall meet many other such pairs, so that we may then use the results established here for them as well.

Supplements and Additional Exercises

Exercise 3.7* (Chebyshev's Inequality) For any mean $\| \ \|^*$, any function f, and any $r > 0$

$$\| [f > r] \|^* \leq \frac{1}{r} \cdot \|f\|^* .$$

Exercise 3.8 Let S be any set, and $\mathcal{E}$ any ring of bounded functions on S that contains the functions of finite carrier. Show that $\| \ \|_u$ satisfies (i) of Theorem 3.5 but is a mean if and only if S is finite.

Exercise 3.9 Suppose S is a locally compact metric space and $\mathcal{E} = C_{00}[S]$ is the lattice ring ring of continuous functions of compact support. Then a function h belongs to $\mathcal{E}^\uparrow$ if and only if it has σ–compact carrier and is lower semicontinuous. (A function h is ***lower semicontinuous*** if the sets $[h > r]$ are open for all $r \in \mathbb{R}$. A subset of S is ***σ–compact*** if it is contained in the countable union of compact sets.)

Exercise 3.10 The Daniell upper integral has a property somewhat sharper than countable subadditivity. Let f_n be an increasing sequence of arbitrary functions — possibly taking the values $\pm\infty$ — with pointwise supremum f. Then

$$\int^* f = \lim_n \int^* f_n \ :$$

"The Daniell upper integral is ***continuous along arbitrary increasing sequences***."

Exercise 3.11* Prove the countable subadditivity of the Daniell mean using its finite subadditivity and the result of Exercise 3.10.

Exercise 3.12 Try to show that $\int^*$ is continuous along arbitrary decreasing sequences: $f_n \downarrow f$ implies $\int^* f_n \downarrow \int^* f$.

Exercise 3.13* Compare the Jordan and Daniell means: $\int^* f \leq \int^\natural f$ and $\|f\|^* \leq \|f\|^\natural$ for all numerical functions f.

Exercise 3.14* (i) Define $\mathcal{E}_\downarrow$ and the lower Daniell integral $\int_*$. (ii) Show that $\int_*$ is positive–homogeneous, increasing, continuous along *decreasing* sequences of $\mathcal{E}_\downarrow$, and countably superadditive. (iii) Show that $\mathcal{E}_\downarrow = -\mathcal{E}^\uparrow$ and $\int_* f = -\int^*(-f)$ and that $\int_*$ is continuous along arbitrary decreasing sequences. (iv) Show that $\int_* f \leq \int^* f \quad \forall f$.

Exercise 3.15 For all positive numerical functions f,

$$\int^* f \, d\lambda = \sup\{ \int^* \phi{\cdot}f \, d\lambda : \phi \in \mathcal{E}_+[\mathbb{R}] , \phi \leq 1 \} ,$$

and consequently for all positive numerical functions f

$$\|f\|^* = \sup\{ \|\phi \cdot f\|^* : \phi \in \mathcal{E}_+[\mathbb{R}] , \phi \leq 1 \} .$$

The Order–Continuous Case. For the remainder of this section $(\mathcal{E}, \int)$ is a positive order–continuous elementary integral (see Definition 2.8). Denote by $\mathcal{E}^\Uparrow$ the collection of all functions on the ambient space that are pointwise suprema of *arbitrary*, not necessarily countable, families of elementary integrands. Define a new upper integral $\int^\bullet$, the ***Daniell–Stone upper integral***, first on functions $h \in \mathcal{E}^\Uparrow$ by

$$\int^\bullet h = \sup\{\int \phi : h \geq \phi \in \mathcal{E}\}$$

and then on arbitrary functions f by

$$\int^\bullet f = \inf\{\int^\bullet h : f \leq h \in \mathcal{E}^\Uparrow\} .$$

Finally, the ***Daniell–Stone mean*** is defined by

$$\|f\|^\bullet = \int^\bullet |f| \;\in \overline{\mathbb{R}}_+ . \tag{3.3}$$

The following exercises develop the properties of these notions in complete analogy with the main body of this section.

Exercise 3.16* (i) $\mathcal{E}^\Uparrow$ is closed under addition, multiplication with positive scalars, and under taking finite infima and arbitrary suprema.
(ii) $\int^\bullet$ is ***continuous along increasingly directed subsets*** of $\mathcal{E}^\Uparrow$; that is to say, for any increasingly directed subset $H \subset \mathcal{E}^\Uparrow$ with pointwise supremum h — which by (i) belongs to $\mathcal{E}^\Uparrow$ —

$$\int^\bullet h = \sup_{h' \in H} \int^\bullet h' .$$

(iii) $h \mapsto \int^\bullet h$ is positive–homogeneous and additive on $\mathcal{E}^\Uparrow$.

Exercise 3.17* $\| \ \|^\bullet$ is a mean and majorizes the elementary integral. Furthermore, $\| \ \|^\bullet$ is order–continuous in the following sense:

Definition 3.18 *A mean* $\| \ \|^*$ *is* ***order–continuous*** *if it is continuous along increasingly directed subsets of* $\mathcal{E}^\Uparrow$.

Exercise 3.19 $\int$ is clearly also σ–additive, so we can construct both $\| \ \|^*$ and $\| \ \|^\bullet$. Show that $\| \ \|^\bullet \leq \| \ \|^*$. Suppose $\mathcal{E}$ contains a countable subset that is uniformly dense; then $\| \ \|^\bullet = \| \ \|^*$.

II.4 Negligible Functions and Sets

Recall the state of affairs: we are facing a mean $\| \ \|^*$ on a ring $\mathcal{E}$ of elementary functions, for instance the Daniell mean of an elementary integral on $\mathcal{E}$. Our goal is the integration theory of $(\mathcal{E}, \| \ \|^*)$, i.e., the investigation of structure of the closure of $\mathcal{E}$ under $\| \ \|^*$. In this section only the solidity and countable subadditivity of the mean $\| \ \|^*$ are exploited.

Definition 4.1 *A function f is called* ***negligible*** *if its mean $\|f\|^*$ is zero. A set is negligible if its indicator function is negligible.*
A property of the points of the underlying set is said to hold ***almost everywhere****, or* ***a.e.*** *for short, if the set of points where it fails to hold is negligible.*

*If we want to emphasize that the definition refers to the solid and countably subadditive seminorm $\| \ \|^*$ we talk about $\| \ \|^*$–negligible functions and $\| \ \|^*$–a.e. convergence, etc. If $\| \ \|^*$ is the Daniell mean $\| \ \|^\lambda$ for Lebesgue measure λ and if we want to emphasize that fact then we use the words "****Lebesgue negligible****."*

Exercise 4.2* Page 27 contains another definition of a Lebesgue negligible set.
(i) Show that it agrees with the present one. (ii) Show that the set $\mathbb{Q}$ of rational numbers is Lebesgue negligible. (iii) Given $\epsilon > 0$, construct a dense open set $U \subset \mathbb{R}$ whose Lebesgue outer measure $\lambda^*(U) \stackrel{\text{def}}{=} \int^* U \, d\lambda$ is less than ϵ.

Here are the permanence properties of negligibility:

Proposition 4.3 *(i) The sum of countably many positive negligible functions is negligible. The union of countably many negligible sets N_n is negligible. Any subset of a negligible set is negligible.*
(ii) A function f is negligible if and only if it vanishes almost everywhere, that is to say, if and only if its carrier $[f \neq 0]$ is negligible.
(iii) If f' and f'' agree almost everywhere then they have the same mean.
(iv) A function with finite mean is finite almost everywhere.

Proof. (i): If f_n, $n = 1, 2, \ldots$, are negligible functions then

$$\Big\| \bigvee_{n=1}^{\infty} |f_n| \Big\|^* \le \Big\| \sum_{n=1}^{\infty} |f_n| \Big\|^* \le \sum_{n=1}^{\infty} \|f_n\|^* = \sum_{n=1}^{\infty} 0 = 0 .$$

The second claim is a particular instance of this, since sets are positive (idempotent) functions (Convention I.2.6). The third claim is immediate from the solidity of $\| \ \|^*$.
(ii): $[f \neq 0](x) \le \sum_{n=1}^{\infty} |f(x)|$. Thus if $\|f\|^* = 0$ then

$$\| \, [f \neq 0] \, \|^* \le \sum_{n=1}^{\infty} \|f\|^* = \sum_{n=1}^{\infty} 0 = 0.$$

Conversely, $|f| \le \sum_{n=1}^{\infty} [f \neq 0]$, so that $\big\| \, [f \neq 0] \, \big\|^* = 0$ implies

$$\|f\|^* \le \sum_{n=1}^{\infty} \| \, [f \neq 0] \, \|^* = \sum_{n=1}^{\infty} 0 = 0.$$

(iii): $\big\| \, [|f' \neq f''] \, \big\|^* = 0$ implies $\|f' - f''\|^* = 0$, and by Exercise I.7.2

$$\Big| \|f'\|^* - \|f''\|^* \Big| \le \|f' - f''\|^* = 0.$$

(iv): $n \cdot [|f| = \infty] \leq |f| \quad \forall\, n \in \mathbb{N}$, and so

$$n \cdot \|\, [|f| = \infty]\, \|^* \leq \|f\|^* \qquad \forall\, n \in \mathbb{N}\,.$$

If $[|f| = \infty]$ is not negligible, $\|f\|^*$ must be infinite. ∎

Functions Defined Almost Everywhere. The only functions of interest for the purpose at hand are, of course, those with finite mean. We should like to argue that the sum of any two of them has finite mean again, in view of the subadditivity of $\|\ \|^*$. A technical difficulty appears: even if f and g have finite mean there may be points x where $f(x) = +\infty$ and $g(x) = -\infty$ or *vice versa*; then $f(x) + g(x)$ is not defined.

The solution to this tiny quandary is to notice that such ambiguities may happen at most in a negligible set of arguments x (see Proposition 4.3 (iv)). We simply extend $\|\ \|^*$ to functions that are defined merely almost everywhere:

Definition 4.4 *Let f be a function defined almost everywhere, that is to say such that the complement of* $\mathrm{dom}(f)$ *is negligible. We set $\|f\|^* = \|f'\|^*$, where f' is any function defined everywhere and coinciding with f almost everywhere in the points where f is defined.*

Part (iii) of the previous proposition shows that this definition is good: it does not matter which function f' we choose to agree a.e. with f; any two will differ negligibly and thus have the same mean. Given two functions f and g with finite mean that are merely almost everywhere defined we define their sum $f + g$ to equal $f(x) + g(x)$ where both $f(x)$ and $g(x)$ are finite. This function is almost everywhere defined, as the set of points where f or g are infinite or not defined is negligible. It is clear how to define the infimum, maximum, product, scalar multiples, etc. of functions that are merely a.e. defined.

From now on, therefore, "***function***" will stand for "almost everywhere defined function" if the context permits it.

Definition 4.5 *An almost everywhere defined function f is said to have finite mean, or to be* ***finite in mean****, if $\|f\|^* < \infty$. $\mathcal{F}^*$ denotes the collection of a.e. defined functions with finite mean. We write $\mathcal{F}^*[\|\ \|^*]$ if we want to stress the point that the mean is $\|\ \|^*$.*

Theorem 4.6 *$\mathcal{F}^*$ is closed under taking finite linear combinations, finite maxima and minima, and under chopping; and $\|\ \|^*$ is a solid and countably subadditive seminorm on $\mathcal{F}^*$. The pair $(\mathcal{F}^*, \|\ \|^*)$ is a complete seminormed space. Of the utmost importance is this fact:* ***Every mean–Cauchy sequence has a subsequence that converges pointwise almost everywhere to a mean–limit.***

Proof. The first statement is left as an exercise. For the remaining two claims let (f_n) be a mean–Cauchy sequence in $\mathcal{F}^*$; that is to say

$$\sup_{m,n\geq N} \|f_m - f_n\|^* \xrightarrow[N\to\infty]{} 0 .$$

For $n = 1, 2, \ldots$ there exists a function f'_n that is everywhere defined and finite and agrees with f_n a.e. Let N_n denote the negligible set of points where f_n is not defined or does not agree with f'_n. There is an increasing sequence (n_k) of indices such that

$$\|f'_n - f'_{n_k}\|^* \leq 2^{-k-1} \text{ for } n \geq n_k .$$

Clearly $\quad \|f'_{n_{k+1}} - f'_{n_k}\|^* \leq 2^{-k}$, so that $g \stackrel{\text{def}}{=} \sum_{k=1}^{\infty} |f'_{n_{k+1}} - f'_{n_k}|$

has finite mean, by countable subadditivity. Therefore the "bad set"

$$B = \bigcup_{n=1}^{\infty} N_n \cup [g = \infty]$$

is negligible (Proposition 4.3 (iv)). If x belongs to the "good set" $G = B^c$ then $g(x) < \infty$ and

$$f(x) = f'_{n_1}(x) + \sum_{k=1}^{\infty} (f'_{n_{k+1}}(x) - f'_{n_k}(x)) = \lim_{k\to\infty} f'_{n_k}(x)$$

exists, since the infinite sum above converges absolutely. Also,

$$\|f - f_{n_K}\|^* = \|f - f'_{n_K}\|^* = \left\|G \cdot (f - f'_{n_K})\right\|^* = \left\|G \cdot \sum_{k=K}^{\infty} f'_{n_{k+1}} - f'_{n_k}\right\|^*$$

$$\leq \left\| \sum_{k=K}^{\infty} |f'_{n_{k+1}} - f'_{n_k}| \right\|^* \leq 2^{-K} \xrightarrow[K\to\infty]{} 0 .$$

Thus $(f'_{n_k})_{k=1}^{\infty}$ converges to f both in mean and pointwise on G. So does the subsequence (f_{n_k}) of the original sequence (f_n). Given $\epsilon > 0$, let K be so large that

$$\| f - f_{n_k}\|^* = \|f - f'_{n_k}\|^* < \epsilon/2 \qquad \text{for } k \geq K$$

and $$\|f_m - f_n\|^* < \epsilon/2 \qquad \text{for } m, n \geq n_K .$$

So if $n \geq n_K$ then $$\| f - f_n \|^* < \|f - f_{n_K}\|^* + \|f_{n_K} - f_n\|^* < \epsilon .$$

This shows that $f_n \xrightarrow[n\to\infty]{} f$ in mean and $f_{n_k} \xrightarrow[n\to\infty]{} f$ both in mean and almost everywhere. ∎

From now on we shall not be so excruciatingly punctilious. If we have to perform algebraic or limit arguments on a sequence of functions that are defined merely almost everywhere we replace without mention every one of them with a function that is defined and finite everywhere and perform the arguments on the resulting sequence; this affects neither the means of the functions nor their convergence in mean or almost everywhere.

Exercise 4.7* Let (f_n) be a mean–convergent sequence with limit f. Any function differing negligibly from f is also a mean limit of (f_n), and any two mean limits of (f_n) differ negligibly.

Exercise 4.8 $\mathcal{F}^*$ is not in general a ring.

II.5 Integrable Functions

In this section $\|\ \|^*$ is a mean on $\mathcal{E}$. For the time being the elementary integrands $\mathcal{E}$ are merely assumed to form a ring of bounded functions on the ambient set. This generality costs a little but comes in handy in connection with Fubini's theorem —see page 128.

Definition 5.1 *An almost everywhere defined function f is called* ***integrable*** *if there is a sequence (ϕ_n) of elementary functions converging in mean to it: $\|f - \phi_n\|^* \to 0$. In other words, f is integrable if it belongs to the mean–closure of $\mathcal{E}$ in $\mathcal{F}^*$. The collection of integrable functions is denoted by $\mathcal{L}^1$.*

If we want to stress the point that the definition refers to the pair $(\mathcal{E}, \|\ \|^*)$ we talk about $(\mathcal{E}, \|\ \|^*)$–integrability and $\mathcal{L}^1[\mathcal{E}, \|\ \|^*]$. If $\|\ \|^*$ is the Daniell mean $\|\ \|^\mu$ for the positive σ–additive elementary integral $\int = \int d\mu$ then the functions in $\mathcal{L}^1$ are called the ***μ-integrable functions****, and we write $\mathcal{L}^1 = \mathcal{L}^1[\mu]$.*

The integrable functions are thus defined with the mean $\|\ \|^*$ just as the Riemann integrable functions were characterized with $\|\ \|^\natural$ in Proposition I.5.1.

The Permanence Properties

As discussed in Section I.6 one does not want to apply the definition to discover whether a given function f is integrable. Rather one establishes once and for all the permanence properties of integrability and checks that f is made up from functions known to be integrable *via* constructions that preserve integrability. It is expedient to do this first, and then to define and examine the integral with these tools at one's disposal.

Theorem 5.2 (Mean Completeness) *(i) If (f_n) is a sequence of integrable functions converging in mean to f then f is integrable.*
(ii) $\mathcal{L}^1$ is complete in mean. Moreover, every mean–Cauchy sequence (f_n) has a subsequence (f_{n_k}) that converges almost everywhere to a mean limit of (f_n).
(iii) If $\mathcal{L}^1 \ni f_n \to f$ in mean and $g = f$ a.e. then $f_n \to g \in \mathcal{L}^1$ in mean as well.

Proof. For (i) copy the proof of Theorem I.6.2, and (ii) is then immediate from Theorem 4.6 on p. 50. (iii) is obvious, just a reminder. ∎

The existence of a pointwise a.e. convergent subsequence of a mean–convergent sequence (f_n) is frequently very helpful in identifying the limit, as we shall presently see. One might be inclined to hope that a mean convergent sequence (f_n), itself, converges pointwise a.e. Alas, this is not so. Here is an example of a sequence (ϕ_n) of *elementary* functions that converges in mean to zero but fails to converge at any single point of $(0,1]$: every natural number n can be written uniquely in the form $n = 2^m + k$, with $m, k \in \mathbb{N}$ and $0 \le k < 2^m$. This being done, let ϕ_n be the indicator function of the interval $(k2^{-m}, (k+1)2^{-m}]$. Clearly $\|\phi_n\|^\lambda = 2^{-n} \to 0$ as $n \to \infty$. Nevertheless, $\phi_n(x)$ takes the values 0 and 1 again and again, for every $x \in (0,1]$. (It might help to draw the graphs of the first few of the ϕ_n.)

Lemma 5.3 *(i) If $\phi \in \mathcal{E}$ then $|\phi|$ is integrable. Therefore a positive integrable function f is the mean limit of positive elementary integrands.*
(ii) Property (M) extends to positive integrable functions: if a sequence (f_n) of them satisfies $\sup_N \|\sum_{n=1}^N f_n\|^* < \infty$ then $\|f_n\|^* \to 0$.

Proof. (i): Set $M = \sup_x |\phi(x)|$, and let (P_n) be a sequence of positive polynomials with zero constant term that converges increasingly to the absolute value function $t \mapsto |t|$ on $[-M, M]$. Such is provided by Corollary I.3.5 on p. 10. Then $\phi_n = P_n \circ \phi \in \mathcal{E}_+$ increases to $|\phi|$. The sequence (ϕ_n) is mean–Cauchy. Indeed, if it were not then there would exist an $\epsilon > 0$ and a subsequence (ϕ_{n_k}) so that the positive elementary integrands $\psi_k \overset{\text{def}}{=} \phi_{n_{k+1}} - \phi_{n_k}$ would have $\|\psi_k\|^* > \epsilon$. This impossible, though, by property (M) of a mean, since $\|\sum_k \psi_k\|^* \le \||\phi|\|^* < \infty$. Theorem 5.2 now provides a mean limit $g \in \mathcal{L}^1$ of (ϕ_n), which must equal $|\phi|$ almost everywhere, since some subsequence of (ϕ_n) converges almost everywhere both to $|\phi|$ and to g. To prove the second claim of (i) let an $\epsilon > 0$ be given. There is an $\phi \in \mathcal{E}$ with $\|f - \phi\|^* < \epsilon/2$. Since $\big|f - |\phi|\big| \le \big|f - \phi\big|$, this implies $\|f - |\phi|\|^* < \epsilon/2$. Then there is a positive $\phi_n \in \mathcal{E}$ with $\||\phi| - \phi_n\|^* < \epsilon/2$, and the subadditivity of $\|\ \|^*$ yields $\|f - \phi_n\|^* < \epsilon$.

(ii): There are *positive* elementary functions ϕ_n with $\|f_n - \phi_n\|^* < 2^{-n}$.

Since

$$\sup_N \Big\| \sum_{n=1}^{N} \phi_n \Big\|^* \leq \sup_N \Big\| \sum_{n=1}^{N} (\phi_n - f_n) \Big\|^* + \sup_N \Big\| \sum_{n=1}^{N} f_n \Big\|^*$$

$$\leq \sum_{n=1}^{\infty} 2^{-n} + \sup_N \Big\| \sum_{n=1}^{N} f_n \Big\|^* < \infty ,$$

we have $\|\phi_n\|^* \xrightarrow[n\to\infty]{} 0$ and thus $\|f_n\|^* \xrightarrow[n\to\infty]{} 0$. ∎

Theorem 5.4 (The Monotone Convergence Theorem or MCT) *Let (f_n) be an increasing or decreasing sequence of integrable functions whose means form a bounded set of reals. Then (f_n) converges to its pointwise limit f in mean.*

Proof. As the sequence $(f_n(x))$ of reals is monotone it has a limit $f(x)$, possibly equal to $\pm\infty$. First the case that the sequence (f_n) is increasing. Then (f_n) is mean–Cauchy. Indeed, assume it were not. There would then exist an $\epsilon > 0$ and a subsequence (f_{n_k}) with $\|f_{n_{k+1}} - f_{n_k}\|^* > \epsilon$. However, this sequence of positive integrable functions satisfies clearly

$$\sup_K \Big\| \sum_{k=1}^{K} f_{n_{k+1}} - f_{n_k} \Big\|^* = \sup_K \|f_{n_K} - f_{n_1}\|^* \leq \|f_{n_1}\|^* + \sup_N \|f_N\|^* < \infty .$$

By Lemma 5.3 (iii), though, $(f_{n_{k+1}} - f_{n_k})$ must converge to zero in mean.

Now that we know that (f_n) is Cauchy, we employ Theorem 5.2: there is a mean–limit f' and a subsequence (f_{n_k}) [10] so that $f_{n_k}(x) \to f'(x)$ for all x outside some negligible set N. For every single x, though, $f_n(x) \to f(x)$. Thus

$$f(x) = \lim_{n\to\infty} f_n(x) = \lim_{k\to\infty} f_{n_k}(x) = f'(x) \text{ for } x \notin N .$$

Hence f is equal almost everywhere to the mean–limit f' and thus is a mean–limit itself.

If (f_n) is decreasing rather than increasing then $(-f_n)$ increases pointwise —and by the above in mean— to $-f$; again $f_n \xrightarrow[n\to\infty]{} f$ in mean. ∎

Theorem 5.5 (Permanence Properties concerning Algebra and Order) *Let f, f' be integrable functions and r a real number. Then $f+f'$, rf, $|f|$, $f \vee f'$, $f \wedge f'$, and $f \wedge 1$ are integrable; so is $f \cdot f'$ if f or f' is bounded.*

In particular, $\mathcal{L}^1$ is a vector lattice closed under chopping, and the bounded functions in $\mathcal{L}^1$ form both a ring and a vector lattice closed under chopping.

[10] Not the same as in the previous argument, which was, after all, shown not to exist.

Proof. If $\mathcal{E}$ is a lattice ring then the proof of Theorem I.6.3 on p. 24 applies word–for–word, provided $\|\ \|^{\natural}$ is replaced everywhere by $\|\ \|^{*}$. Nothing was used in that proof beyond the subadditivity and solidity of the seminorm and the fact that $\mathcal{E}$ is a lattice ring.

If $\mathcal{E}$ is merely known to be a ring then $\mathcal{L}^1$ is a vector space as the very first argument in the proof of Theorem 6.1 shows. The fact that the product of two integrable functions is integrable provided one of them is bounded(!) can again be taken literally from Theorem I.6.3. This leaves the permance properties concerning the order. Assume then f is integrable. There exists a sequence (ϕ_n) of elementary integrands converging in mean to f. Since $\big|\,|f|-|\phi_n|\,\big| \le |f-\phi_n|$, $\big\|\,|f|-|\phi_n|\,\big\|^* \le \big\|\,|f-\phi_n|\,\big\|^* \xrightarrow[n\to\infty]{} 0$, so $|f|$ is integrable, inasmuch as the $|\phi_n|$ are —see Lemma 5.3 (i) and Theorem 5.2 (i).

By Exercise I.2.13 on p. 8, $\mathcal{L}^1$ is a vector lattice.

To see that it is closed under chopping, the only thing not yet proved, let $f \in \mathcal{L}^1$. In order to show that $f\wedge 1 \in \mathcal{L}^1$ it suffices to show that $(f\wedge 1)_+ = f_+ \wedge 1$ is integrable, because $(f\wedge 1)_- = f\wedge 0$ certainly is, and $f\wedge 1$ is the sum of these two functions. In other words, we may assume that f is positive. Now if $\phi \in \mathcal{E}_+$ then $\phi\wedge 1$ is the uniform limit of an increasing sequence of positive elementary integrands (Exercise I.3.18) and thus is integrable. If $f \ge 0$ is merely integrable, we approximate it in mean by a sequence (ϕ_n) of *positive* elementary integrands — so that $\phi_n \wedge 1$ is integrable — and use the solidity of $\|\ \|^*$ on the inequality $|f\wedge 1 - \phi_n \wedge 1| \le |f-\phi_n|$ to show that $f\wedge 1$ is integrable as well. ∎

The permanence properties concerning order and algebraic operations are thus not any better than for the Jordan norm $\|\ \|^{\natural}$. The ones concerning limits are much superior, though, and permit the construction of vast numbers of new integrable functions. The best is yet to come; compare the following result with Theorem I.6.5 on p. 26.

Theorem 5.6 (The Dominated Convergence Theorem or DCT) *Let (f_n) be a sequence of integrable functions and assume that*

(i) $f_n \xrightarrow[n\to\infty]{} f$ *pointwise almost everywhere and*
(ii) $|f_n| \le g$ *for some function g with finite mean and all indices n.*

Then (f_n) converges to f in mean, and consequently f is integrable.

Proof. As in the proof of the Monotone Convergence Theorem we begin by showing that the sequence (f_n) is Cauchy. To this end consider the positive function

$$g_N = \sup\{|f_n - f_m| : m,n \ge N\} = \lim_{K\to\infty} \bigvee_{m,n=N}^{K} |f_n - f_m| \quad \le 2g.$$

By Theorem 5.5 and the Monotone Convergence Theorem, g_N is integrable. Moreover, $g_N(x)$ converges decreasingly to zero at all points x at which $f_n(x)$ converges, that is, almost everywhere. By Exercise 5.7,

$$\|g_N\|^* \xrightarrow[N\to\infty]{} 0 .$$

Now $\|f_n - f_m\|^* \le \|g_N\|^*$ for $m, n \ge N$, so (f_n) is indeed Cauchy in mean. By Theorem 5.2, the sequence has a mean limit f' and a subsequence (f_{n_k}) that converges pointwise a.e. to f'. Since (f_{n_k}) also converges to f a.e., f and f' agree almost everywhere, namely, at all points x at which both $(f_n(x))$ and $(f_{n_k}(x))$ converge. Thus $\|f_n - f\|^* = \|f_n - f'\|^* \xrightarrow[n\to\infty]{} 0$. ∎

The Dominated Convergence Theorem is central. Most other results in integration theory follow from it. It is false without some domination condition such as (ii), as the following example shows. Let ϕ_n be the elementary function equal to n on $(0, 1/n]$ and zero elsewhere; the sequence (ϕ_n) converges to zero at every single point, yet $\|\phi_n\|^\lambda = \int \phi_n(x)\,dx = 1$ for all n. However, a slightly weaker condition, uniform integrability, suffices (Proposition 5.20). The assumption of pointwise a.e. convergence in Theorem 5.6 and Proposition 5.20 can also be replaced with a weaker condition, convergence in measure. These matters are deferred until later (Corollary III.4.12 on p. 84).

Supplements and Additional Exercises

For the while, $\| \;\|^*$ is still an arbitrary mean on a ring $\mathcal{E}$, not necessarily the Daniell mean for some σ–additive elementary integral.

Exercise 5.7* Show in detail that the conclusion of Theorem 5.4 continues to hold if the f_n are merely defined a.e. and are increasing or decreasing merely a.e.

Exercise 5.8* A function f is integrable if and only if there is a sequence (ϕ_n) of elementary functions with $\sum_{n=1}^\infty \|\phi_n\|^* < \infty$ and $f = \sum_{n=1}^\infty \phi_n$ almost everywhere.

Exercise 5.9* A function $h \in \mathcal{E}^\uparrow$ is integrable if and only if $\|h\|^* < \infty$. The mean is continuous along increasing sequences of $\mathcal{E}^\uparrow$.

Exercise 5.10 The collection $\mathcal{L}^1_b$ of bounded integrable functions is a lattice ring closed under chopping and mean–dense in $\mathcal{L}^1$. So is the collection $\mathcal{L}^1_{b0} = (\mathcal{L}^1_b)_0$ of $\mathcal{L}^1_b$–confined functions in $\mathcal{L}^1_b$.

Exercise 5.11 Suppose $\mathcal{E}$ is a lattice ring. The collection $\mathcal{E}_{00}$ of $\mathcal{E}$–confined functions in $\mathcal{E}$ then is a self–confining lattice ring uniformly dense in $\mathcal{E}$ and mean–dense in $\mathcal{L}^1$.

Exercise 5.12 (i) $\mathcal{L}^1$ is not in general a ring; (ii) in fact, it is a ring precisely if it is finite–dimensional, and then it is an algebra. (iii) Discuss when this can happen.

Exercise 5.13* (Fatou's lemma) The following is an easy but very useful corollary of the Monotone Convergence Theorem: let (f_n) be a sequence of positive integrable functions. Then

$$\|\liminf_{n\to\infty} f_n\|^* \le \liminf_{n\to\infty} \|f_n\|^* ;$$

and if $\liminf_{n\to\infty} \|f_n\|^*$ is finite then $\liminf_{n\to\infty} f_n$ is integrable.

Exercise 5.14 The composition $\gamma \circ f$ of a continuous function γ with an integrable function f is integrable, provided there is a function $h \in \mathcal{F}^*$ with $|\gamma \circ f| \leq h$ a.e.

Exercise 5.15 A function $g : \mathbb{R} \to \mathbb{R}$ is called a ***Lipschitz function*** if there is a constant K such that $|g(s) - g(t)| \leq K|s - t|$ for all $s, t \in \mathbb{R}$. Show that the composition $g \circ f$ of a Lipschitz function g that vanishes at zero with a bounded integrable function f is integrable.

Exercise 5.16 (On Order–continuous Means) Assume the mean $\| \ \|$ is order–continuous, say $\| \ \|$ is the Daniell–Stone mean $\| \ \|^\bullet$ of an order–continuous elementary integral (see page 48). Prove: if Φ is an increasingly directed or decreasingly directed family of elementary integrands with $\sup_{\phi \in \Phi} \|\phi\|^\bullet < \infty$ then $\sup \Phi$ or $\inf \Phi$, respectively, is integrable and $\Phi \to \lim \Phi$ in mean. The conclusion continues to hold if Φ is an increasingly directed subset of $\mathcal{E}^\Uparrow$ with $\sup_{h \in \Phi} \|h\|^\bullet < \infty$.

Uniform Integrability. To discuss this notion, let us first introduce ***order intervals***: Let $g, g' \in \mathcal{L}^1$. The order interval $[g, g']$ is defined by

$$[g, g'] = \{f \in \mathcal{L}^1 : g \leq f \leq g'\}.$$

The domination condition (ii) in the Dominated Convergence Theorem amounts to requiring that the f_n all lie in some fixed order interval $[-g, g]$.

If f is any function in $\mathcal{L}^1$ there is a function $f_g^{g'}$ in $[g, g']$ closest to f in mean, to wit:

$$f_g^{g'} = g \vee f \wedge g' .$$

It is obtained by chopping f below with g and above with g'. It is advisable to draw a picture.

Exercise 5.17 Show that $\|f - f_g^{g'}\|^* \leq \|f - f'\|^* \quad \forall f' \in [g, g']$.

Definition 5.18 *A collection $\mathcal{S}$ of integrable functions is called* ***uniformly integrable*** *if for every $\epsilon > 0$ there is an order interval $[g, g'] \subset \mathcal{L}^1$ such that the mean distance from f to $[g, g']$:*

$$\inf_{f' \in [g, g']} \|f - f'\|^* = \|f - f_g^{g'}\|^* = \|f - (g \vee f \wedge g')\|^*$$

is less than ϵ for all $f \in \mathcal{S}$.

Exercise 5.19 $\mathcal{S}$ is uniformly integrable if and only if there is, for every $\epsilon > 0$, an elementary function ψ such that for all $f \in \mathcal{S}$

$$\|f - (-\psi \vee f \wedge \psi)\|^* < \epsilon .$$

Uniform integrability can replace condition (ii) of the Dominated Convergence Theorem; and this much domination is also necessary. We leave the proof of this fact as an exercise:

Proposition 5.20 *Let (f_n) be a sequence of integrable functions converging pointwise a.e. to some function f. Then (f_n) converges in mean to f if and only if the family $\{f_n : n \in \mathbb{N}\}$ is uniformly integrable.*

II.6 Extending the Integral

In this section we assume that the mean $\| \ \|^*$ majorizes the elementary integral:

$$\left|\int \phi\right| \le \|\phi\|^* .$$

This holds, for instance, if $\| \ \|^*$ is the Daniell mean for $\int$, but there are other instances (Section V.2). The elementary integrands are assumed to form a ring $\mathcal{E}$.

The ***extended integral*** is now defined exactly as in Proposition I.5.1 on p. 21, with $\| \ \|^*$ replacing the Jordan mean[11] $\| \ \|^\natural$. Namely, if f is a $\| \ \|^*$–integrable function there is a sequence (ϕ_n) of elementary integrands converging in mean to f, and the integral of f is defined as

$$\int f = \lim \int \phi_n .$$

Why is the integral well–defined?[12] First, since $\| \ \|^*$ majorizes the integral,

$$\begin{aligned}\left|\int \phi_n - \int \phi_m\right| &= \left|\int (\phi_n - \phi_m)\right| \\ &\le \|\phi_n - \phi_m\|^* \le \|\phi_n - f\|^* + \|f - \phi_m\|^* \xrightarrow[m,n\to\infty]{} 0.\end{aligned}$$

Thus $(\int \phi_n)$ is a Cauchy sequence of reals and does have a limit. Next, if (ϕ'_n) is a second sequence of elementary functions converging in mean to f then

$$\|\phi_n - \phi'_n\|^* \le \|\phi_n - f\|^* + \|f - \phi'_n\|^* \xrightarrow[n\to\infty]{} 0 ;$$

so $|\int \phi_n - \int \phi'_n| = |\int \phi_n - \phi'_n| \to 0$, and $(\int \phi_n)$ and $(\int \phi'_n)$ have the same limit: the integral is well defined.

Here are some basic properties of the integral:

Theorem 6.1 *(i) The extended integral is linear and increasing; that is to say, for any $f, g \in \mathcal{L}^1$ and $r, s \in \mathbb{R}$*

$$\int (rf + sg) = r\int f + s\int g ,$$

and

$$f \le g \implies \int f \le \int g .$$

[11] which is no mean, of course, see page 46

[12] If you did Exercise I.7.17 on p. 29 you know.

(ii) The integral is still majorized by the mean: for $f \in \mathcal{L}^1$, $\left|\int f\right| \le \|f\|^*$. *If* $\|\ \|^*$ *agrees with* $\int$ *on the positive elementary integrands then*

$$\left|\int f\,\right| \le \int |f| = \int^* |f| = \|f\|^* \qquad \forall f \in \mathcal{L}^1 .$$

(iii) If the sequence (f_n) *of integrable functions converges in mean to* f *then* f *is integrable, and* $\int f_n \xrightarrow[n\to\infty]{} \int f$.
(iv) Suppose the sequence (f_n) *of integrable functions increases or decreases a.e. to a function* f *and* $\lim \int f_n$ *is finite* **or** (f_n) *converges a.e. to* f *and* $\int^* \sup |f_n|$ *is finite. Then* $f_n \to f$ *in mean,* f *is integrable, and* $\int f_n \to \int f$.

Proof. The first statement is an instance of Exercise I.7.17: Let $(\phi_n), (\psi_n)$ be sequences in $\mathcal{E}$ converging in mean to $f, g \in \mathcal{L}^1$, respectively. Then $(r\phi_n + s\psi_n)$ converges to $rf + sg$ in mean, since

$$\left\|(rf+sg) - (r\phi_n + s\psi_n)\right\|^* \le |r| \cdot \left\|f - \phi_n\right\|^* + |s| \cdot \left\|g - \psi_n\right\|^* \xrightarrow[n\to\infty]{} 0 .$$

Thus

$$\int (rf+sg) = \lim_n \int (r\phi_n + s\psi_n) = \lim_n \left(r\int \phi_n + s\int \psi_n \right) = r\int f + s\int g .$$

This shows that the integral is linear. By taking differences the monotonicity is seen to be equivalent to

$$f \ge 0 \Longrightarrow \int f \ge 0 \quad \forall f \in \mathcal{L}^1 .$$

By Lemma 5.3, there are positive elementary functions converging in mean to f. The limit of their integrals, $\int f$, is then positive as well.
(ii): Let (ϕ_n) be a sequence of elementary functions with $\left\|f - \phi_n\right\|^* \xrightarrow[n\to\infty]{} 0$. As $\Big||f| - |\phi_n|\Big| \le |f - \phi_n|$, $\left\||f| - |\phi_n|\right\|^* \xrightarrow[n\to\infty]{} 0$ and $\int |f| = \lim \int \phi_n$. The elementary inequality $\left|\int \phi_n\right| \le \int |\phi_n|$ produces

$$\left|\int f\,\right| = \lim_n \left|\int \phi_n\right| \le \lim_n \int |\phi_n| \le \lim_n \|\phi_n\|^* .$$

Now $\|\phi_n\|^* \xrightarrow[n\to\infty]{} \|f\|^*$ by Exercise I.7.2 on p. 28, and the desired inequalities follow. (iii) is evident and so is (iv) when $\|\ \|^*$ is the Daniell mean for $\int$; the general case is left as a (non–trivial) exercise. ▮

Exercise 6.2* Assume not only that $\|\ \|^*$ majorizes the elementary integral $\int$, but that it is, specifically, the Daniell mean $\|\ \|^\mu$ for the σ–additive elementary integral $\int d\mu$. Show that a function f is then $(\mathcal{E}, \|\ \|^*)$–integrable if and only if

$$-\infty < \int_* f = \int^* f < +\infty \tag{6.1}$$

(see Exercise 3.14). In this case the common value $\int^* f = \int_* f$ is the integral. Compare this with Proposition I.5.1 on p. 21.

Remark 6.3 The Dominated Convergence Theorem is often stated as follows: *If the sequence (f_n) of integrable functions is dominated and converges to the function f a.e. then f is integrable and $\int f_n \xrightarrow[n\to\infty]{} \int f$.*
This is evident, since $f_n \xrightarrow[n\to\infty]{} f$ in mean and the integral is continuous in mean. This statement seems genuinely weaker than the DCT. However, it is not really, given some argument: Prove the following generalization of the DCT: Let (g_n) be a sequence of integrable functions converging almost everywhere to an integrable function g and assume that $\lim_n \int g_n = \int g$. If (f_n) is a sequence of integrable functions converging to f almost everywhere and if $|f_n| \leq g_n \quad \forall n$ then (f_n) converges to f in mean.

The Lebesgue Integral.

If $\| \ \|^*$ is the Daniell mean $\| \ \|^\lambda$ for the usual integral on the line we write $\mathcal{L}^1[\lambda]$ or $\mathcal{L}^1[\mathbb{R}]$ for $\mathcal{L}^1$ and talk about ***Lebesgue integrable functions***. The integral of an integrable function f is then called its ***Lebesgue integral*** and written as $\int f\, d\lambda$ or $\int f(x)\, dx$. We shall even mark the upper integral and write $\int^* f(x)\, \lambda(dx) = \int^* f(x)\, dx$ instead of just $\int^*$.

How does Lebesgue's integral compare with Riemann's?

Proposition 6.4 *A Riemann integrable function is Lebesgue integrable, and its Riemann integral coincides with its Lebesgue integral.*

Proof. From Exercise 3.13, $\| \ \|^* \leq \| \ \|^\natural$. So if (ϕ_n) is a sequence of elementary functions that converges to the Riemann integrable function f in Jordan mean then it converges in Daniell–mean:

$$\|f - \phi_n\|^* \leq \|f - \phi_n\|^\natural \xrightarrow[n\to\infty]{} 0\,.$$

The integral in either sense is the limit of the sequence $(\int \phi_n)$ of reals. ∎

Supplements and Additional Exercises

The mean $\| \ \|^*$ is now Daniell's mean $\| \ \|^\mu$ for the σ–additive elementary integral $\int \, d\mu$.

Exercise 6.5 For any $f \in \mathcal{F}^*[\| \ \|^*]$ there is an integrable function h with $|f| \leq h$.

Exercise 6.6 A function f with finite upper integral $\int^* f$ is majorized by an integrable function $\overline{f}$ with $\int^* f = \int \overline{f}$. Such $\overline{f}$ is called an integrable ***upper envelope*** of f. A function f with finite lower integral $\int_* f$ majorizes an integrable function $\underline{f}$ with $\int_* f = \int \underline{f}$. Such $\underline{f}$ is called a integrable ***lower envelope*** for f. If f is a set then $\overline{f}$ and $\underline{f}$ can be chosen to be sets.

Exercise 6.7 $\int^*(f+f') = \int^* f + \int^* f'$ if one of f, f' is integrable.

Exercise 6.8 Let $\| \ \|$ be a mean on $\mathcal{E}[\mathbb{R}]$ having $\|\phi\| = \int \phi$ for all positive elementary integrands. Then $\|f\| \le \|f\|^*$ for all $f : S \to \overline{\mathbb{R}}$.

Exercise 6.9 Suppose someone has somehow constructed a vector space $\mathcal{L}^*$ of functions containing $\mathcal{E}$ and a linear and positive extension $\int$ of the elementary integral to $\mathcal{L}^*$ such that the MCT and DCT hold; that is to say, if (f_n) is a sequence in $\mathcal{L}^*$ that either increases with $\sup \int f_n < \infty$ or that converges pointwise being dominated by a function in $\mathcal{L}^*$, then the limit function f belongs to $\mathcal{L}^*$ and $\int f = \lim \int f_n$. Show that then there is a mean $\| \ \|$ majorizing the elementary integral and such that $\mathcal{L}^*$ is the closure of $\mathcal{E}$ with respect to it.

Exercise 6.10* $C_{00}[\mathbb{R}]$ is dense in $\mathcal{L}^1[\mathbb{R}]$. $(\mathcal{L}^1[\lambda], \| \ \|^*)$ is separable.

Exercise 6.11* Lebesgue integral is ***translation invariant*** — the meaning of this will become clear during the exercise. For $a \in \mathbb{R}$ and any function f on the line denote by f_a the translated function $x \mapsto f(x-a)$. Then for all $a \in \mathbb{R}$, $\phi \in \mathcal{E}$, and $f : \mathbb{R} \to \mathbb{R}$
(i) $\phi \in \mathcal{E}[\mathbb{R}]$ if and only if $\phi_a \in \mathcal{E}[\mathbb{R}]$: the collection $\mathcal{E}[\mathbb{R}]$ is translation invariant.
(ii) $\int \phi_a(x)\,\lambda(dx) = \int \phi(x)\,\lambda(dx)$: the elementary integral is translation invariant.
(iii) $\int^* f_a(x)\,\lambda(dx) = \int^* f(x)\,\lambda(dx)$: the upper integral is translation invariant.
(iv) $f \in \mathcal{L}^1[\lambda]$ if and only if $f_a \in \mathcal{L}^1[\lambda]$, and then $\int f(x)\,\lambda(dx) = \int f_a(x)\,\lambda(dx)$: the class of integrable functions and the extended integral are translation invariant.

Exercise 6.12* Let f be a Lebesgue integrable function such that $\int f(x)\phi(x)dx$ vanishes for every continuous function ϕ of compact support. Show that $f = 0$ a.e.
Can we draw the same conclusion if $\int f(x)\phi(x)\,dx = 0$ for every $\phi \in C^\infty_{00}[\mathbb{R}]$?

II.7 Integrable Sets

Again, the arguments of the section apply to any mean.

According to our Convention I.2.6 on p. 6, sets are identified with their indicator functions: *A set is integrable if its indicator function is integrable.* The permanence properties of the class of integrable sets are immediate from those of the class of integrable functions:

Proposition 7.1 *The union and set difference of two integrable sets are integrable. The intersection of a countable family of integrable sets is integrable. The union of a countable family of integrable sets is integrable provided it is contained in an integrable set C.*

Proof. Let $A_1, A_2, \ldots$ be a countable family of integrable sets. Then

$$A_1 \cup A_2 = A_1 \vee A_2 \ ,$$
$$A_1 \backslash A_2 = A_1 - (A_1 \wedge A_2) \ ,$$

$$\bigcap_{n=1}^{\infty} A_n = \bigwedge_{n=1}^{\infty} A_n = \lim_{N\to\infty} \bigwedge_{n=1}^{N} A_n \ ,$$

and
$$\bigcup_{n=1}^{\infty} A_n = C - \bigwedge_{n=1}^{\infty} (C - A_n) \ ,$$

in the sense that in every line the set on the left has — or *is*, see Convention I.2.6—the indicator function on the right, which is integrable by the permanence properties of Section 5. ▮

Some Notation will facilitate the discussion of integrable sets and their connection with integrable functions. Recall from Proposition 2.13 on p. 39 that a collection of subsets of some set is a ring of sets if it is closed under taking finite unions and relative complements — and then under taking finite intersections. A ring of sets that is closed under taking countable intersections is called a ***δ–ring***. Proposition 7.1 can thus be expressed by saying that the integrable sets form a δ–ring.
A finite linear combination of integrable sets is clearly integrable. It is an integrable function that takes only finitely many values, every nonzero one in an integrable set. Such functions are called ***integrable simple functions***. In other words, the integrable simple functions are the step functions over the δ–ring of integrable sets.
A set A is called ***σ–finite*** if it can be covered by countably many integrable sets. In many instances the whole ambient set is σ–finite. In that case we say that the ***mean is σ–finite***. If the mean is Daniell's mean for some elementary integral and is σ–finite, then we also say the elementary integral or measure is σ–finite. This is the same as saying the ambient space is the union of countable many integrable sets. Most means are σ–finite. The only example of a mean that is not and can still be considered a natural example is that of the Daniell mean of a Radon measure on a locally compact space so huge that it cannot be covered by countably many compact sets.

Exercise 7.2* If there is a sequence of elementary integrands whose pointwise supremum is strictly positive then any mean is σ–finite. Thus Lebesgue integral on the line or on $\mathbb{R}^n$ is σ–finite. Conversely, if *the Daniell mean* of $\int$ on $\mathcal{E}$ is σ–finite then there exists a sequence of elementary integrands $\phi_n \in \mathcal{E}_+$ with $\sup_n \phi_n = 1$.

Exercise 7.3 If $\| \ \|^* \neq 0$ is σ–finite then there is a countable collection of disjoint non–negligible integrable sets whose union is the ambient space.

Exercise 7.4* If the mean is σ–finite then any collection $\mathcal{M}$ of mutually disjoint non–negligible integrable sets is at most countable.

Recall from Proposition 2.13 on p. 39 that a collection of subsets of some set is an algebra of sets if it is a ring and contains the whole ambient space. An algebra of sets that is closed under taking countable intersections is automatically closed under taking countable unions and is called a ***σ–algebra***. The integrable sets do not, in general, form a σ–algebra — the line $\mathbb{R}$, for instance, is not Lebesgue integrable. If they do, then we say the mean is ***finite*** or ***totally finite***. If the mean is Daniell's mean for some elementary integral and is totally finite, then we also say the elementary integral or measure is totally finite.

Let us now investigate the relation of the integrable sets to the integrable functions.

Proposition 7.5 *Let f be an integrable function. (i) For any strictly positive real r, $[f > r]$, $[f \geq r]$, $[f < -r]$, and $[f \leq -r]$ are integrable sets. (ii) f is the limit almost everywhere and in mean of a sequence (f_n) of simple integrable functions with $|f_n| \leq |f|$. (iii) f has σ-finite carrier.*

Proof. For the first claim, note that

$$[f > 1] = \lim_{n\to\infty} 1 \wedge \big(n(f - f \wedge 1)\big)$$

is integrable: the functions $f_n = 1 \wedge \big(n(f - f \wedge 1)\big)$ are integrable by Theorem 5.5 and are dominated by $|f|$. By the DCT, their limit is integrable. This limit is zero at any point x where $f(x) \leq 1$ and 1 at any point x where $f(x) > 1$; in other words, it is (the indicator function of) the set $[f > 1]$, which is therefore integrable. *This is the first and only place where we use the fact that $\mathcal{L}^1$ is closed under chopping.*
The set $[f > r]$ equals $[f/r > 1]$ and is thus integrable as well. Finally,

$$[f \geq r] = \Big\{\bigcap [f > r - 1/n] : \mathbb{N} \ni n > 1/r\Big\}$$

is integrable, and $[f < -r] = [-f > r]$ and $[f \leq -r] = [-f \geq r]$.

For (ii), let f_n be the simple integrable function

$$\begin{aligned} f_n = &\sum_{k=1}^{2^{2n}} k2^{-n} \cdot [k2^{-n} < f \leq (k+1)2^{-n}] \\ &- \sum_{k=1}^{2^{2n}} k2^{-n} \cdot [-k2^{-n} > f \geq -(k+1)2^{-n}] . \end{aligned} \tag{7.1}$$

By (i), the sets

$$[k2^{-n} < f \leq (k+1)2^{-n}] = [f > k2^{-n}] \setminus [f > (k+1)2^{-n}]$$

are integrable if $k \neq 0$. Thus f_n, being a linear combination of integrable functions, is integrable. Now (f_n) converges pointwise to f and is dominated by $|f|$, and the claim follows.

As for (iii), the carrier $[f \neq 0]$ of f equals the union of the integrable sets $[|f| > 1/n]$. ∎

Remark 7.6 Suppose the mean $\| \|^*$ is Daniell's mean $\| \|^\mu$ for some elementary integral $\int d\mu$. Let us write $\mu(A)$ for $\int A\, d\mu$ when A is an integrable set. According to Equation (7.1), the integral of an integrable function f is the limit of sums of the form

$$I_n = \sum_{1 \le |k| \le 2^n} k2^{-n} \mu(A_n) , \tag{7.2}$$

where $\quad A_n \stackrel{\text{def}}{=} \{x : f(x) \text{ lies between } k2^{-n} \text{ and } (k+1)2^{-n}\} .$

Proposition 7.5 can then be read as follows. Instead of chopping the domain into little pieces as is done in the Calculus to approximate the integral of f by a Riemann sum, subdivide its range into intervals of length 2^{-n} and replace the Rieman sum by the sums (7.2). It is sometimes said that this "transition from the x–axis to the y–axis" is Lebesgue's great contribution; this makes things sound too easy to do him justice. There is, after all, the not–so–small matter of attaching a measure to the sets A_n.

Exercise 7.7* Let f_n be a sequence of integrable functions, all vanishing off the same integrable set A and converging uniformly to f. Then $f_n \to f$ in mean and f is integrable.

Exercise 7.8 Let f be a positive function. The following are equivalent:
(i) f is integrable; (ii) there exists an increasing sequence (s_n) of simple integrable functions with $f = \sup s_n$ a.e. and $\sup \int s_n < \infty$. In this case the previous supremum is the integral of f, provided $\| \|^*$ is the Daniell mean of $\int$; (iii) there exists a sequence (s_n) of simple integrable functions with $f = \sum s_n$ a.e. and $\sum \int s_n < \infty$. In this case the previous sum is the integral of f if $\| \|^*$ is the Daniell mean of $\int$.

Assume henceforth that an elementary integral $\int = \int d\mu$ is given and that the mean majorizes the integral:

$$\left| \int \phi \, d\mu \right| \le \|\phi\|^* \quad \forall \phi \in \mathcal{E} . \quad \text{Then} \quad \left| \int f(x)\, \mu(dx) \right| \le \|f\|^* \quad \forall f \in \mathcal{L}^1 .$$

The integral of an integrable set A, $\int A(x)\, \mu(dx)$, is usually denoted by $\mu(A)$. Let f be an integrable function. An integrable set A being a bounded function, the product $A \cdot f = 1_A \cdot f$ is integrable. Its integral is usually denoted by

$$\int_A f \quad \text{or} \quad \int_A f\, d\mu \quad \text{or} \quad \int_A f(s)\mu(ds) .$$

Exercise 7.9* **(A Borel–Cantelli Lemma)** Given a sequence (A_n) of integrable sets that are all contained in some other integrable set, let A denote the set of points that lie in infinitely many of them. A is integrable. If $\sum_n \int A_n < \infty$ then A is negligible.

Exercise 7.10* **(The Riesz Representation Theorem)** The restriction of the integral to the integrable sets is a positive σ–additive measure. The step functions over this ring, the simple integrable functions, are dense in $\mathcal{L}^1$.

Exercise 7.11 Exercise 7.10 produces a σ–additive set function μ on the δ–ring $\mathcal{R} \subset \mathcal{L}^1$ of integrable sets. We are in the situation of page 38. We would then produce a elementary integral by linear extension and then extend this, for instance by constructing its Daniell mean, or by using statement (ii) of Proposition 7.5 as the definition if integrability, to a new class of integrable functions. Show that either way we end up right back at $\mathcal{L}^1$: much ado with nothing to show for it.

The remainder of the section deals specifically with the Lebesgue integral. We write $\int_A f(x)\,\lambda(dx)$ for $\int_A f$ in order to emphasize that it is Lebesgue measure we are integrating with. The δ–ring of Lebesgue integrable sets is denoted by $\mathcal{A}[\lambda]$. The integral $\int 1_A(x)\,\lambda(dx)$ of a Lebesgue integrable set is traditionally written as $\lambda(A)$ and called the ***Lebesgue measure*** of A. Similarly, the upper integral of (the indicator function of) a subset A of the line is denoted by $\lambda^*(A)$ and called the ***outer Lebesgue measure*** of A.

Proposition 7.12 (Regularity of Lebesgue Measure) *Recall that an open subset of the line is the union of at most countably many mutually disjoint open intervals (a_i, b_i), among which may or may not be one or both of the infinite intervals $(-\infty, b)$ and (a, ∞). For the purpose of this proposition let us call this representation of an open set its **canonical representation**.*
(i) The outer Lebesgue measure of an open set is the sum of the lengths of the intervals in its canonical representation.
(ii) A bounded open subset of the line is Lebesgue integrable, and its Lebesgue measure is then the sum of the lengths of the intervals in its canonical representation.
(iii) A compact subset of the line is Lebesgue integrable.

Now let A be any Lebesgue integrable set, and let $\epsilon > 0$.
*(iv) There exists an open integrable set U containing A with $\lambda(U\setminus A) < \epsilon$. This fact is called **outer regularity** of Lebesgue measure.*
*(v) There exists a compact set $K \subset A$ with $\lambda(A\setminus K) < \epsilon$. This fact is called **inner regularity** of Lebesgue measure.*

Additional Exercises

Exercise 7.13 Every σ–additive measure on the line is inner and outer regular.

The remaining exercises deal specifically with the Lebesgue integral.

Exercise 7.14 (See Exercise 3.15) Show that

$$\int^* f\,d\lambda = \sup\{\int_A^* f : A \text{ integrable}\}$$

for all positive numerical functions f, and consequently

$$\|f\|^* = \sup\{\|A\cdot f\|^* : A \text{ integrable}\}\,.$$

Exercise 7.15 Let A be a subset of the line, and let $a \in \mathbb{R}$. Show that the right translate $A + a = \{x + a : x \in A\}$ has the same outer measure as A: $\lambda^*(A + a) = \int^* 1_{A+a} = \int^* A = \lambda^*(A)$. Show that if A is Lebesgue integrable then so is $A + a$ and has the same measure. [Hint: $A + a = A_a$ in the notation of Exercise 6.11 on p. 61.]

The following project permits the reader to check her understanding of the material presentend so far.

Project 7.16 *For $\phi \in C_{00}[\mathbb{R}]$ let $\int \phi$ denote its Riemann integral. The triple $(\mathbb{R}, C_{00}[\mathbb{R}], \int)$ has the same properties as the triple $(\mathbb{R}, \mathcal{E}[\mathbb{R}], \int)$ we have been considering so far. Namely, $C_{00}[\mathbb{R}]$ is a lattice ring of bounded functions, and $\int : C_{00}[\mathbb{R}] \to \mathbb{R}$ is linear, positive, and σ–additive (Use Lemma I.3.1). Define the upper integral for this triple as in Section 3, replacing $\mathcal{E}$ everywhere by $C_{00}[\mathbb{R}]$. Define the corresponding mean, the integrable functions, the extended integral, and produce the permanence properties of the past sections.*
Then show that the integrable functions so obtained are exactly the Lebesgue integrable ones, and that the integrals coincide.

II.8 Example of a Non–Integrable Function

Let us try to think of a non–integrable function f on the line. There is $f(x) = x^2$, which can't be integrable because $\|f\|^* = \infty$: "f is too big." We understand this obstruction to integrability very well. Let us accept it and look for a "small" function f that is not integrable. Let us try to find one that is bounded and of compact support, say, so that its mean is finite. This is not easy: how would we describe such a function? How does one usually describe a function? Well, one starts with simple ones, polynomials or step functions, say, and produces more complicated ones as algebraic or order combinations or as limits of these. The permanence properties of the class $\mathcal{L}^1$ are so good, though, that we have no chance at producing our non–integrable function that way. We might be led to the conjecture that every function f with $\|f\|^* < \infty$ is integrable. Alas, this is not true. This section exhibits an example of a bounded set $B \subset \mathbb{R}$ that is not Lebesgue integrable. However, this is not easy and, in fact, requires a logical device not openly employed so far, the axiom of choice.

We start the construction of B by defining an equivalence relation $\sim$ on $\mathbb{R}$ by

$$r \sim r' \iff r - r' \in \mathbb{Q} .$$

Every equivalence class is countable, and there are uncountably many of them. Every equivalence class clearly has an element between -1 and 1 in it. Using the axiom of choice, we pick from every equivalence class a number between -1 and 1 to make the "bad" set B.

(i) The sets $B+q$ and $B+q'$ are disjoint if $q \neq q' \in \mathbb{Q}$. Indeed, if their intersection has a number $b+q = b'+q'$ in it, $b, b' \in B$, $q, q' \in \mathbb{Q}$, then $b \sim b'$ and consequently $b = b'$, since only one representative of every equivalence class is in B. In that case q equals q', and the two classes are identical.

Let us enumerate the rationals in $(-2,2)$: $\mathbb{Q} \cap (-2,2) = \{q_1, q_2, \ldots\}$ and let B_n be B translated by q_n: $B_n = B + q_n$, $n \in \mathbb{N}$. The sets B_n are countable in number, mutually disjoint, and contained in the interval $(-3,3)$.

(ii) The interval $(-1,1)$ *is contained in the union of the* B_n. Indeed, if $r \in (-1,1)$ then r is in one of the eqivalence classes and differs from the representative chosen from this class by a rational q which cannot be bigger than 2 in absolute value. q is one of the q_n and consequently $r \in B_n$. From the *countable subadditivity* of the upper integral we deduce now

$$2 = \int^* (-1,1) \leq \int^* \bigcup_{n=1}^{\infty} B_n \leq \sum_{n=1}^{\infty} \int^* B_n .$$

Thus one of the numbers $\int^* B_n$ must be strictly positive. By Exercise 7.15, all the B_n have the same upper integral, and so

$$\int^* B = \int^* B_n > 0 \qquad \forall\, n \in \mathbb{N} .$$

Now assume, by way of contradiction, that B is Lebesgue integrable. Then so are all the B_n (Exercise 7.15), and the countable *additivity* of the measure on integrable sets yields

$$\int^* (\bigcup_{n=1}^{\infty} B_n) = \sum_{n=1}^{\infty} \int B_n = \sum_{n=1}^{\infty} \int B = +\infty.$$

This is impossible, since $\bigcup_{n=1}^{\infty} B_n \subset (-3,3)$ and thus $\lambda^*(\bigcup_{n=1}^{\infty} B_n) \leq 6$. This contradiction shows that B cannot be Lebesgue integrable. ∎

Exercise 8.1 Construct a decreasing sequence f_n of functions with pointwise infimum zero such that $\int^* f_n \not\to 0$.

Chapter III

Measurability

Recall the example of a non–integrable function given in Section II.8. It was (the indicator function of) a bounded set B. The reason this function is not integrable clearly is its wild behaviour: it jumps at every point. But then there is $\mathbb{Q}$, the indicator function of the rationals; it also jumps at every point, yet it *is* integrable. The difference between these two wildly jumping functions is that the latter is negligible; there is an arbitrarily small set off which $\mathbb{Q}$ is actually constant —see Exercise II.4.2 on p. 49. This observation leads to the conjecture that the local behaviour of an integrable function is this: it is allowed to behave wildly on small sets but has to be somewhat smooth elsewhere. One would not conjecture that an integrable function is constant on large sets, but, perhaps, that it is as smooth as an elementary function there. Once conjectured, this is not hard to prove — see Observation 1.1 below. This local behaviour of the integrable functions is called measurability.

It was Littlewood who made this observation and related ones. Because of their importance they are known as *Littlewood's Principles*: integrable functions are "nearly elementary," integrable sets are "nearly elementary sets," and measurable functions are "locally nearly elementary."

It is a hallmark of the most insightful results in mathematics that they get turned into definitions. This is an instance. After detailing Littlewood's Principles in the first section we shall define a function to be measurable if it is, on arbitrarily large sets, as smooth as an elementary function.

Measurability describes precisely the "local structure" of the integrable functions: a function f is integrable if and only if it is measurable and its mean is finite. Its permanence properties are superior to those of integrability in that no domination is required for the limit. The main use of this notion is this: Suppose we are confronted with a function f and want to determine whether it is integrable. Since measurability is preserved under algebraic and order operations and under limits of sequences, it is frequently quite easy to see from the way f is given that it is measurable; we prove in some other way that its mean is finite and conclude that it is integrable. This is often much preferable to checking that the domination condition of the Dominated Convergence Theorem holds in the various steps that comprise the definition of f.

The arguments are quite general in nature; they apply to any mean $\| \ \|^*$ for a ring $\mathcal{E}$ of bounded functions.

K. Bichteler, *Integration – A Functional Approach*, Modern Birkhäuser Classics,
DOI 10.1007/978-3-0348-0055-6_3, © Birkhäuser Verlag 1998

III.1 Littlewood's Principles

Observation 1.1 *Let f be $(\mathcal{E}, \| \ \|^*)$–integrable and $\epsilon > 0$. There exists a set $U \in \mathcal{E}^\uparrow$ with $\|U\|^* < \epsilon$ and a function f' almost everywhere equal to f that is, on the whole complement of U, the uniform limit of elementary functions. Therefore (Corollary I.3.13) f itself coincides with a function of $\overline{\mathcal{E}}$ on a set whose complement has mean less than ϵ.*
We boldly paraphrase this by saying "an integrable function is largely from $\overline{\mathcal{E}}$."

Proof. The proof of Theorem II.4.6 on p. 50 provides a sequence of elementary functions ϕ_k with $\|\phi_k\|^* < 2^{-k}$ for $k = 1, 2, \ldots$ so that $f = \sum_{k=0}^{\infty} \phi_k$ in mean and almost everywhere. We set f' equal to f where this sum converges, equal to zero elsewhere. Consider the elementary integrands

$$\psi_K \stackrel{\text{def}}{=} \sum_{1 \le k < K} k \cdot \phi_k \quad \text{and set} \quad g' \stackrel{\text{def}}{=} \lim_{K \to \infty} \psi_K = \sum_{k=1}^{\infty} k \cdot |\phi_k| \,.$$

Since
$$\|g' - \psi_K\|^* \le \sum_{K \le k} k 2^{-k} \xrightarrow[K \to \infty]{} 0 \,,$$

g' is integrable. Observe next that for any $M, K \in \mathbb{N}$

$$\|[g' > M]\|^* \le \|g'/M\|^* \le M^{-1} \sum_{k<K} \|k \cdot \phi_k\|^* + \sum_{k \ge K} k 2^{-k} \,.$$

By choosing first K and then M sufficiently large, we can have $\|[g' > M]\|^* < \epsilon$. $U \stackrel{\text{def}}{=} [g' > M]$ is a set in $\mathcal{E}^\uparrow$: it equals the supremum of $1 \wedge \big(n \cdot (\psi_K - (\psi_K \wedge M))\big)$ over n, K. On its complement U^c we have $g' \le M$ and thus for all $K \in \mathbb{N}$

$$|f' - \sum_{k<K} \phi_k| \le \sum_{k \ge K} |\phi_k| \le K^{-1} \sum_{k \ge K} k \cdot |\phi_k| \le g'/K \le M/K \,,$$

showing that on this set the convergence of $\sum_k \phi_k$ to f is uniform. The set of the second claim is $U^c \cup [f \ne f']$. This argument does not even use the assumption that $\| \ \|^*$ be a mean; solidity and countable subadditivity suffice! ∎

Observation 1.2 *Let (f_n) be a sequence of integrable functions and assume that $f_n(x)$ converges to $f(x)$ for almost all x in the integrable set A. For every $\epsilon > 0$ there is then an integrable subset $A_0 \subset A$ with $\|A \setminus A_0\|^* < \epsilon$ such that (f_n) converges to f* ***uniformly*** *on A_0.*

To paraphrase this, there are "arbitrarily large" subsets of A on which the convergence is actually uniform, or "*almost everywhere convergence is largely uniform.*"

Proof. By Proposition II.7.5 on p. 63,

$$A^r_{m,n} \stackrel{\text{def}}{=} A \cap \left[|f_n - f_m| > \frac{1}{r} \right]$$

is an integrable set for $r, n, m = 1, 2, \ldots$, and then so is the set (see Proposition II.7.1 on p. 61)

$$B^r_p \stackrel{\text{def}}{=} \bigcup_{m,n \geq p} A^r_{m,n} = A \cap \bigcup_{m,n \geq p} \left[|f_n - f_m| > \frac{1}{r} \right] .$$

As p increases, B^r_p decreases, and the intersection $\bigcap_p B^r_p$ is contained in the negligible set of points where (f_n) does not converge (I.2.5). Thus, by the Monotone Convergence Theorem,

$$\lim_{p \to \infty} \|B^r_p\|^* = 0 .$$

For every $r \in \mathbb{N}$ there is therefore a natural number $p(r)$ such that

$$\|B^r_{p(r)}\|^* < 2^{-r} \epsilon .$$

Set
$$B = \bigcup_r B^r_{p(r)} \qquad \text{and} \quad A_0 = A \backslash B .$$

It is evident that $\|A \backslash A_0\|^* = \|B\|^* < \epsilon$. It is left to be shown that (f_n) converges uniformly on A_0. To this end let $\delta > 0$ be given. We let $N = p(r)$, where r is chosen so that $1/r < \delta$. Now if x is any point in A_0 and $m, n \geq N$ then x is not in the "bad set" $B^r_{p(r)}$, and therefore $|f_n(x) - f_m(x)| \leq 1/r < \delta$. Thus $|f(x) - f_n(x)| \leq \delta$ for all $x \in A_0$ and $n \geq N$. ∎

Considering a Lebesgue integrable function f note this consequence: due to Theorem II.4.6 f is $\| \ \|^\lambda$–almost everywhere the pointwise limit of elementary functions. By Observation 1.2, f is on arbitrarily large integrable sets the *uniform* limit of elementary functions. Moreover, by Theorem II.4.6 and Exercise II.6.10, f is a.e. the pointwise limit of *uniformly continuous* functions. It is thus on arbitrarily large integrable sets the uniform limit of uniformly continuous functions and so is uniformly continuous on them. Furthermore, every Lebesgue–integrable set contains an arbitrarily large compact subset, by Proposition II.7.12 on p. 65, and on compact sets continuity and uniform continuity coincide. To summarize:

Corollary 1.3 *Let f b a Lebesgue integrable function, A a Lebesgue integrable set, and $\epsilon > 0$.*
(i) There exists an integrable subset $A_0 \subset A$ with $\lambda(A \backslash A_0) < \epsilon$ such that f is, on A_0, the uniform limit of elementary functions.
(ii) There exists an integrable subset $A_0 \subset A$ with $\lambda(A \backslash A_0) < \epsilon$ such that f is uniformly continuous on A_0.

(iii) There exists a compact subset $K \subset A$ with $\lambda(A \setminus K) < \epsilon$ such that the restriction of f to K is continuous.

The upshot of all this is that a Lebesgue integrable function is rather regular on large sets.

Definition of Measurability

Recall that we are searching for the local structure of the integrable functions, to be called measurability. Corollary 1.3 offers the answer: A function f should be called measurable if it behaves as described there. Which of the three possibilities shall we choose? The second one is particularly appealing, exhibiting as it does the local smoothness in a succinct and intuitive way; however, the first one is the easiest to express for an *arbitrary mean* $\| \ \|^*$, so we settle for it [1]:

Definition 1.4 *Let A be an integrable set. A real–valued almost everywhere defined function f is called* **measurable on** *A if for every $\epsilon > 0$ there is an integrable subset A_0 of A with $\|A \setminus A_0\|^* < \epsilon$ on which f is the uniform limit of elementary functions, that is to say*[2]*, on which f coincides with a function in the uniform closure $\overline{\mathcal{E}}$.*
A real–valued a.e. defined function f is called **measurable** *if it is measurable on every integrable set.*
If there is need to stress that this definition refers to the pair $(\mathcal{E}, \| \ \|^*)$ then f will be called $(\mathcal{E}, \| \ \|^*)$–measurable. If $\| \ \|^*$ is the Daniell mean for Lebesgue measure on the line or on $\mathbb{R}^n$ we talk about **Lebesgue measurability***.*

Supplements and Additional Exercises

Exercise 1.5* The constant function 1 is measurable.
Exercise 1.6 Let $\| \ \|^*$ be the Daniell mean for counting measure (see Example II.2.22). Then every function is $\| \ \|^*$–measurable.

An Alternative Definition of Measurability. The following contemplations are for the reader with a good background in topology; in particular, he must be at ease with the general Stone–Weierstrass Theorem I.3.25 on p. 16.

Definition 1.4 can be read as follows: f *is measurable if and only if it is $\mathcal{E}$–uniformly continuous on arbitrarily large sets.* This statement could have served as the definition of measurability. Doing that has two advantages: First, it makes sense also for functions f that take values in some other uniform space, for instance, in a metric space or a Banach space—this is true also for statement (ii) of Corollary 1.3.

---

[1] See, however, Definition 1.7 below.
[2] See Corollary I.3.13 on p. 14.

Second, it does not refer to the topology of the line and so it makes sense also when the space S on which one wishes to integrate has *a priori* no topology. It may even happen that S has a relevant topology but that it is too cumbersome to carry through all the arguments.

Definition 1.7 *A function f with values in a uniform space is called measurable if for every integrable set A and $\epsilon > 0$ there is an integrable subset $A_0 \subset A$ with $\|A \backslash A_0\|^* < \epsilon$ on which f is $\mathcal{E}$–uniformly continuous. To paraphrase as above: f is measurable if it is "**largely $\mathcal{E}$–uniformly continuous.**"*

Exercise 1.8 A real–valued function is $\mathcal{E}$–uniformly continuous on A_0 if and only if it agrees there with a function of $\overline{\mathcal{E}}$.

If we don't like the $\mathcal{E}$–uniformity in the alternate Definition 1.7 we don't have to use it. Let $\mathcal{C}$ be any collection of everywhere defined bounded integrable functions and assume that the ring $\mathcal{E}'$ generated by it is dense in $\mathcal{L}^1$. It is not hard to see that the $\mathcal{E}'$–uniformity and the $\mathcal{C}$–uniformity are equivalent, in the sense that they have the same uniformly continuous functions. By Corollary 3.3, the statement "*f is measurable if it is $\mathcal{C}$–uniformly continuous on arbitrarily large sets*" is true and can serve as the definition of measurability:

Exercise 1.9 Use this definition with $\mathcal{C}$ the collection of functions $t \mapsto t \cdot [-n, n]$ to prove (ii) of Corollary 1.3. More generally, show that a function f with values in some uniform space is Lebesgue measurable if and only if for every integrable set A and $\epsilon > 0$ there is an integrable subset $A_0 \subset A$ on which f is uniformly continuous (with respect to the usual distance $|\ |$ on $\mathbb{R}$).

III.2 The Permanence Properties

Again $\|\ \|^*$ is an arbitrary mean on a ring $\mathcal{E}$ of bounded functions.

The notion 1.4 of measurability is rather intuitive, describing as it does a large degree of smoothness. We have yet to show that it captures exactly the local structure of the integrable functions. This we refer to later sections, as it is more expeditious to establish the permance properties of measurability first. They are excellent, as we shall presently see. The task is facilitated greatly by the following little result.

Lemma 2.1 *Let A be an integrable set and (f_n) a sequence of functions that are measurable on A. For every $\epsilon > 0$ there exists an integrable subset A_0 of A with $\|A \backslash A_0\|^* \leq \epsilon$ such that everyone of the f_n agrees on A_0 with a function in $\overline{\mathcal{E}}$.*

Proof. Let $A_1 \subset A$ be integrable with $\|A \backslash A_1\|^* < \epsilon 2^{-1}$ and so that, on A_1, f_1 agrees with a function $\overline{\phi}_1 \in \overline{\mathcal{E}}$. Next let $A_2 \subset A_1$ be integrable with $\|A_1 \backslash A_2\|^* <$

$\epsilon 2^{-2}$ and so that, on A_2, f_2 agrees with a function $\overline{\phi}_2 \in \overline{\mathcal{E}}$. Continue by induction, and set $A_0 = \bigcap_{n=1}^{\infty} A_n$. Then A_0 is integrable by Proposition II.7.1,

$$\|A \backslash A_0\|^* = \Big\|(A \backslash A_1) \cup \bigcup_{i>1} (A_i \backslash A_{i-1})\Big\|^* \leq \sum \epsilon 2^{-n} = \epsilon$$

due to the countable subadditivity of $\| \ \|^*$, and every f_n coincides on A_0 with a function in $\overline{\mathcal{E}}$, inasmuch as it does so on the larger set A_n. ∎

Lemma 2.1 renders the proof of the permanence properties of measurability rather simple. We start with limits:

Theorem 2.2 (Egoroff's Theorem) *Let (f_n) be a sequence of measurable functions and assume that (f_n) converges $\| \ \|^*$–a.e. to a function f. Then f is measurable. Moreover, for every integrable set A and $\epsilon > 0$ there is an integrable subset A_0 of A with $\|A \backslash A_0\|^* < \epsilon$ on which (f_n) converges uniformly to f — we paraphrase this behaviour by saying "**(f_n) converges uniformly on arbitrarily large sets** or even (f_n) **converges largely uniformly.**"*

Proof. Let A and $\epsilon > 0$ be given. There is an integrable set $A_1 \subset A$ with $\|A \backslash A_1\|^* < \epsilon/2$ such that, on it, everyone of the f_n coincides with some function in $\overline{\mathcal{E}}$. Thus $f_n \cdot A_1$ is integrable for every $n \in \mathbb{N}$ (why?). These functions converge to $f \cdot A_1$ pointwise a.e., so there is an integrable set $A_0 \subset A_1$ with $\|A_1 \backslash A_0\|^* < \epsilon/2$ on which the convergence is uniform (Observation 1.2). Clearly $\|A \backslash A_0\|^* < \epsilon$, and f agrees on A_0 with a function in $\overline{\mathcal{E}}$ (Corollary I.3.13). ∎

Theorem 2.3 *Finite algebraic and order combinations of real–valued measurable functions are measurable. The composition $\psi \circ f$ of a continuous function ψ with a measurable function f is measurable (see Proposition 5.16 on p. 91 for more).*

Proof. Let f, f' be measurable. Given A integrable and $\epsilon > 0$, we find $A_0 \subset A$ integrable with $\|A \backslash A_0\|^* < \epsilon$ and functions $\overline{\phi}, \overline{\phi}' \in \overline{\mathcal{E}}$ that agree on A_0 with f, f', respectively. Then $f * f'$ agrees on A_0 with $\overline{\phi} * \overline{\phi}' \in \overline{\mathcal{E}}$. For $*$ read $+, -, \cdot, \vee, \wedge$. The functions $|f|, f \wedge 1$ coincide on A_0 with $|\overline{\phi}|, \overline{\phi} \wedge 1$, respectively.
For the last statement let (p_n) be a sequence of polynomials that converge pointwise to ψ (Exercise I.3.15 (iii)). Then $p_n \circ f$ is measurable by the previous argument, and then so is the pointwise limit $\psi \circ f$ of this sequence. The following consequence is left as an exercise: ∎

Corollary 2.4 *Let (f_n) be a sequence of real–valued measurable functions. Then the functions* $\inf f_n$, $\sup f_n$, $\liminf f_n$, $\limsup f_n$ *are measurable as well, provided they are finite almost everywhere.*

Additional Exercises

Exercise 2.5 Deduce Theorem 2.3 from the following fact: if $f_1, \ldots, f_n$ are measurable functions and $\phi : \mathbb{R}^n \to \mathbb{R}$ is continuous then $\phi(f_1, \ldots, f_n)$ is measurable. Prove this fact.

Exercise 2.6 The definition of measurability does not permit us to call measurable as smooth a function as the constant ∞, the reason being that we don't have a notion of uniform approximation in $\overline{\mathbb{R}}$. This can easily be remedied: A natural metric on $\overline{\mathbb{R}}$ is the arctan metric d of Exercise I.2.4. Use this to define the notion of measurability of a $\overline{\mathbb{R}}$–valued function, and check that all the previous results continue to hold, and to improve Corollary 2.4.

The following exercises are for the reader with sufficient background in topology. They use the "proper" Definition 1.7 of measurability.

Exercise 2.7 Prove Egoroff's theorem for a sequence (f_n) of functions with values in a metric space (E, d).

Exercise 2.8 Let f be a measurable function with values in a uniform space E and ϕ a real–valued function on E. Assume (a) that ϕ is uniformly continuous or (b) that E is complete and ϕ continuous. Show that $\phi \circ f$ is measurable. Show that the same is true if ϕ takes values in some uniform space.

Exercise 2.9 Let (E, d) be a complete metric space and assume there is a countable collection $\mathcal{C}$ of uniformly continuous functions that separates the points of E. Show that a function with values in E is measurable iff the following two conditions are satisfied: (a) $\gamma \circ f$ is measurable for all $\gamma \in \mathcal{C}$ and (b) f has locally nearly relatively compact range; that is to say, for every integrable set A and $\epsilon > 0$ there is an integrable subset A_0 of A with $\|A - A_0\|^* < \epsilon$ such that $f(A_0)$ is relatively compact in (E, d).

III.3 The Integrability Criterion

Let us now show that the notion of measurability captures exactly the degree of smoothness of the integrable functions.

Theorem 3.1 (The Integrability Criterion) *A function f is integrable if and only if it is measurable and has finite mean and σ–finite carrier*[3].

Proof. An integrable function has finite mean and is measurable, being almost everywhere the pointwise limit of a sequence of elementary functions (Theorem II.5.2 on p. 53). By part (iii) of Proposition II.7.5 on p. 63 it has σ–finite carrier. The three conditions are thus necessary.

For the sufficiency assume to start with that f vanishes off an integrable set A. There are then integrable subsets A_k of A with $\|A \setminus A_k\|^* < 2^{-k}$ on which f

[3] If the mean $\| \ \|^*$ is Daniell's mean then this follows, of course, from $\|f\|^* < \infty$.

is the uniform limit of elementary functions. Due to Exercise II.7.7 $f_k = f \cdot A_k$ is integrable. If x belongs to $\bigcap_{k>K} A_k$ then $f_k(x) = f(x)$ for $k > K$; thus $f_k(x) \xrightarrow[k\to\infty]{} f(x)$ if $x \in \bigcup_K \bigcap_{k>K} A_k$. Now this set has negligible complement in A. In other words, (f_k) converges to f a.e. Since this sequence is also dominated, by $|f| \in \mathcal{F}^*$, f itself is integrable.
In the general case use a countable collection $\{A_1, A_2, \ldots\}$ of integrable sets that cover the carrier of f. Replacing if necessary A_n by $\bigcup_{k\leq n} A_k$ the A_n can be chosen to increase with n. The functions $f \cdot A_n$ are clearly measurable, so they are integrable by what was shown above. The sequence (f_n) converges to f at every single point and is dominated by $|f| \in \mathcal{F}^*$. We conlude with the Dominated Convergence Theorem that f is integrable. ∎

Exercise 3.2* If the mean is σ–finite or is Daniell's mean or is maximal (Section 6) then a function of finite mean automatically has σ–finite carrier

Corollary 3.3 *Let $\mathcal{D}$ be any collection of integrable functions that is mean–dense in $\mathcal{L}^1$. A real–valued function f is measurable if and only if for every integrable set A and $\epsilon > 0$ there is an integrable subset A_0 of A with $\|A \setminus A_0\|^* < \epsilon$ on which f is the uniform limit of functions in $\mathcal{D}$.*

Proof. Suppose f is measurable. Given an integrable set A and $\epsilon > 0$ we find $A_0' \subset A$ integrable with $\|A \setminus A_0'\|^* < \epsilon/2$ such that f coincides on A_0' with a function in $\overline{\mathcal{E}}$. Then $f \cdot A_0'$ is integrable and is therefore the limit in mean and a.e. of a sequence (d_n) in $\mathcal{D}$. By Observation 1.2, there is an integrable set $A_0 \subset A_0'$ with $\|A_0' \setminus A_0\|^* < \epsilon/2$ on which (d_n) converges uniformly. Clearly $\|A \setminus A_0\|^* < \epsilon$, and f is, on A_0, the uniform limit of functions in $\mathcal{D}$.
Conversely, assume "f is largely the uniform limit of functions in $\mathcal{D}$." Let an integrable set A and $\epsilon > 0$ be given. Let $A_0' \subset A$ be integrable with $\|A \setminus A_0'\|^* < \epsilon/2$ and (d_n) a sequence in $\mathcal{D}$ converging uniformly on A_0' to f. Then $f \cdot A_0'$, being the dominated limit of $(d_n \cdot A_0')$ is integrable and thus is the limit a.e. of a sequence (ϕ_n) of elementary functions. By Observation 1.2, (ϕ_n) converges uniformly to f on some integrable subset A_0 of A_0' with $\|A_0' \setminus A_0\|^* < \epsilon/2$. Clearly $\|A \setminus A_0\|^* < \epsilon$. ∎

Exercise 3.4 Redo Exercises II.5.15 and II.5.14.

The Localization Principle

The following result shows nicely that measurability describes, indeed, the local structure of the integrable functions. In its consequence, the recommended Exercise 3.10 on p. 78, this point becomes particularly clear.

Theorem 3.5 *(i) A function measurable on the integrable set A is measurable on every integrable subset of A. (ii) A function measurable on the integrable sets A^1, A^2 is measurable on their union. (iii) If f is measurable on the integrable sets $A^1, A^2, \ldots$ then it is measurable on every integrable subset of their union.*

Proof. (i) is utterly obvious. (ii): $A^2 \backslash A^1$ is measurable and f is measurable on this set, by (i). We may thus assume that A^1 and A^2 are disjoint. Let us write $B = A^1 \cup A^2$. Given $\epsilon > 0$ we find an integrable subset B_1 of B with $\|B \backslash B_1\|^* < \epsilon/2$ on which both (the indicator functions of) A^1, A^2 coincide with functions $\overline{\alpha}^1, \overline{\alpha}^2$, respectively. Now we find $A_0^i \subset A^i \cap B_1$ with $\|(A^i \cap B_1) \backslash A_0^i\|^* < \epsilon/4$ on which f agrees with $\overline{\phi}^i \in \overline{\mathcal{E}}$, $i = 1, 2$, and set $B_0 = A_0^1 \cup A_0^2$. Clearly $\|B \backslash B_0\|^* < \epsilon$. On B_0, f agrees with $\overline{\alpha}^1 \overline{\phi}^1 + \overline{\alpha}^2 \overline{\phi}^2 \in \overline{\mathcal{E}}$. Thus f is measurable on $A^1 \cup A^2$.
(iii): Let $A \subset \bigcup_{n=1}^{\infty} A^n$ be integrable and let $\epsilon > 0$ be given. There is an integer N such that $\|A \setminus \bigcup_{n=1}^{N} (A \cap A^n)\|^* < \epsilon/2$. By (i), f is measurable on every one of the sets $A \cap A^n$, by (ii) on their union $\bigcup_{n=1}^{N} (A \cap A^n)$. There is an integrable subset A_0 of that union with $\| \bigcup_{n=1}^{N} (A \cap A^n) \setminus A_0\|^* < \epsilon/2$ on which f coincides with a function from $\overline{\mathcal{E}}$. Clearly $\|A \backslash A_0\|^* < \epsilon$. ∎

Corollary 3.6 *A function is measurable if and only if it is measurable on every set of the form $[\phi > r]$, $\phi \in \mathcal{E}$, $r > 0$.*

Proof. Let A be integrable. There are elementary integrands $0 \le \phi_n \le 1$ converging in mean to A. The sets $[\phi_n > 1/r]$ cover A almost everywhere. ∎

Absolute Continuity of Means Consider now two means $\| \ \|^\flat$, $\| \ \|^*$ on $\mathcal{E}$. We say that $\| \ \|^\flat$ is ***absolutely continuous*** with respect to $\| \ \|^*$ and write $\| \ \|^\flat \ll \| \ \|^*$ if every $\| \ \|^*$–negligible set is $\| \ \|^\flat$–negligible.

Corollary 3.7 *If $\| \ \|^\flat \ll \| \ \|^*$ then every $\| \ \|^*$–measurable function is $\| \ \|^\flat$–measurable. Two means with the same negligible sets thus have the same measurable functions.*

Proof. It suffices to show that a $\| \ \|^*$–measurable function f is $\| \ \|^\flat$–measurable on sets of the form $A = [\phi > r]$, which are integrable for both means. If $G \subset A$ is $\| \ \|^*$–integrable then there exists a sequence $\phi_n \in \mathcal{E}$ converging in $\| \ \|^*$–mean and $\| \ \|^*$–almost everywhere to G. So does the sequence of functions $\psi_n = A \cdot \phi_n$ of functions which are integrable for both means. Since this latter sequence converges $\| \ \|^\flat$–almost everywhere to G, this set is $\| \ \|^\flat$–integrable. Let now $\mathcal{G}$ be a maximal collection of mutually disjoint integrable (for both means) and non–negligible subsets of A on which f is the restriction of a function in $\overline{\mathcal{E}}$. Then $\mathcal{G}$ is countable and exhausts A $\| \ \|^*$–almost everywhere. Indeed, if $A \setminus \bigcup \mathcal{G}$

were not negligible, one could adjoin to $\mathcal{G}$ an integrable non–negligible subset of A on which f agrees with an element of $\overline{\mathcal{E}}$, in contradiction to the maximality of $\mathcal{G}$. Therefore $A \cdot f = \sum\{f \cdot G : G \in \mathcal{G}\}$ $\| \ \|^*$–almost everywhere, and thus $\| \ \|^\flat$–almost everywhere: f is $\| \ \|^\flat$–measurable on A. ∎

Supplements and Additional Exercises

Exercise 3.8 The mean of a positive measurable function f is the supremum of the means of the positive simple integrable functions majorized by f.

Exercise 3.9 Let $\mathcal{C}$ be a collection of integrable set so that every integrable set is a.e. contained in the union of a countable subcollection of $\mathcal{C}$ — we say $\mathcal{C}$ is an ***adequate cover***. Show that a function f is measurable if and only if it is measurable on every set of $\mathcal{C}$. Show that a σ–finite mean has an adequate cover.

Exercise 3.10 Let f be a function on the line and $\| \ \|^*$ a mean for $\mathcal{E}[\mathbb{R}]$. Show that f is measurable iff every point $x \in \mathbb{R}$ has a neighborhood on which f is measurable.

Complex Valued Integrands. The order has played such an important rôle in the preceding arguments that one might have misgivings about handling complex–valued functions $f = u + iv$. There is really no problem: one simply declares such a function to be integrable if both its real part u and its imaginary part v are, and in that case one defines the integral by

$$\int f = \int u + i \int v .$$

The Dominated Convergence Theorem holds if the absolute value is understood in the complex sense: $|f| = \sqrt{u^2 + v^2}$.

Exercise 3.11 If $f = u + iv$ is integrable then so is its absolute value $|f|$.

Exercise 3.12* Here is an equivalent way to subsume complex–valued function into the theory: Define $\| \ \|^*$ on arbitrary complex–valued functions f by

$$\|f\|^* = \Big\| |f| \Big\|^* .$$

Then $\| \ \|^*$ is a seminorm on the space $\mathcal{F}^*_{\mathbb{C}}$ of almost everywhere defined complex–valued functions f with $\|f\|^* < \infty$. This space is a complex vector space complete under $\| \ \|^*$, and every Cauchy sequence has an almost every convergent subsequence.
Next let $\mathcal{E} \otimes \mathbb{C}$ denote the collection of all functions ϕ that can be written in the form

$$\phi = \sum_{i=1}^{I} c_i \cdot \phi_i \qquad c_i \in \mathbb{C}, \ \phi_i \in \mathcal{E} .$$

(This is the same as saying $\phi = \rho + i\sigma$, with $\rho, \sigma \in \mathcal{E}$.) These functions are reasonably called the *complex–valued elementary integrands*. They form a complex vector space contained in $\mathcal{F}^*_{\mathbb{C}}$. A complex–valued function f is integrable in the sense above precisely if it belongs to the $\| \ \|^*$–closure $\mathcal{L}^1_{\mathbb{C}}$ of $\mathcal{E}\otimes\mathbb{C}$ in $\mathcal{F}^*_{\mathbb{C}}$. The Dominated Convergence Theorem continues to hold.

Define a complex–valued function to be measurable if for every integrable set A and $\epsilon > 0$ there is an integrable subset $A_0 \subset A$ with $\|A - A_0\|^* < \epsilon$ on which f is the uniform limit of complex–valued elementary integrands, or, which is the same, on which f agrees with a function from the uniform closure $\overline{\mathcal{E}\otimes\mathbb{C}} = \overline{\mathcal{E}}\otimes\mathbb{C}$. Prove the permanence properties: the measurable complex–valued functions form an algebra; Egoroff's theorem und the integrability Theorem 3.1 hold. Show that f is measurable if and only if its real and imaginary parts are.

Exercise 3.13 (The Riemann–Lebesgue Lemma) Take the measure space to be $(-\pi,\pi]$ equipped with the step functions $\mathcal{E}(-\pi,\pi]$ and their usual integral. If f is an integrable function, then its Fourier coefficients a_n, b_n, given by Equation (II.2), converge to zero as $n \to \infty$.

Exercise 3.14 (The Riemann–Lebesgue Lemma) The *Fourier transform* $\mathcal{F}f$ of a Lebesgue integrable function $f : \mathbb{R} \to \mathbb{R}$ is defined by

$$\mathcal{F}f(\alpha) = \int_{-\infty}^{\infty} f(x)e^{i\alpha x}\,dx, \ \alpha \in \mathbb{R} .$$

Show that the integrand is actually integrable. Show that $\mathcal{F}f$ is a continuous bounded function of α that vanishes at infinity.

Exercise 3.15 The Fourier transform is one–to–one in the sense that two integrable functions with the same Fourier transform are a.e. equal.

Banach Space Valued Integrands. The construction of Exercise 3.12 can easily be extended to functions with values in a Banach space $(E, \| \ \|_E)$. To start with we extend the definition of $\| \ \|^*$ to such a function f by

$$\|f\|^* = \Big\| \|f\|_E \Big\|^*$$

and define $\mathcal{F}^*_E$ as the collection of all E–valued almost everywhere defined functions on which $\| \ \|^*$ is finite.

Exercise 3.16 $(\mathcal{F}^*_E, \| \ \|^*)$ is a complete seminormed vector space, and every Cauchy sequence has an almost every convergent subsequence.

Next let $\mathcal{E}\otimes E$ denote the collection of all functions ϕ that can be written as

$$\phi = \sum_{i=1}^{I} \xi_i \cdot \phi_i , \qquad \xi_i \in E , \ \phi_i \in \mathcal{E} . \tag{3.1}$$

These functions are reasonably called the E–valued elementary integrands. They form a vector space contained in $\mathcal{F}^*_E$. Define the E–valued integrable functions $\mathcal{L}^1_E$ as the $\| \ \|^*$–closure of $\mathcal{E}\otimes E$ in $\mathcal{F}^*_E$. The functions in $\mathcal{L}^1_E$ are called ***Bochner integrable***.

Exercise 3.17 They form a vector space complete under the seminorm $\| \ \|^*$. If $f \in \mathcal{L}^1_E$ then $\|f\|_E \in \mathcal{L}^1_{\mathbb{R}}$. The Dominated Convergence Theorem continues to hold.
Define a E–valued function to be measurable if for every integrable set A and $\epsilon > 0$ there is an integrable subset $A_0 \subset A$ with $\|A - A_0\|^* < \epsilon$ on which f is the uniform limit of E–valued elementary integrands. Prove the permanence properties: the measurable E–valued functions form a vector space; the product of one of them with a real–valued measurable function is another one of them; if E is a complex Banach space then the product of one of them with a complex valued measurable function is another one of them; Egoroff's theorem und the integrability Theorem 3.1 hold.

Assume now that $\| \ \|^*$ majorizes the real–valued elementary integral $\int$, for instance that it is its Daniell mean. For ϕ as in (3.1) define the integral by

$$\int \phi = \sum_{i=1}^{I} \xi_i \cdot \int \phi_i \,. \tag{3.2}$$

This is a linear map $\int : \mathcal{E} \otimes E \to E$ — provided it is well defined. We do have to worry about whether $\int \phi$ is independent of the particular choice of the representation (3.1) of ϕ. The next exercises will guide the reader through a proof of the independence and define the extension of the integral to a linear map $\int : \mathcal{L}^1_E \to E$.

Exercise 3.18 Suppose $\mathcal{E}$ consists of step functions over a ring $\mathcal{R}$ of sets. Then $\mathcal{E} \otimes E$ are exactly the E–valued step functions over $\mathcal{R}$: for every $\phi \in \mathcal{E} \otimes E$

$$\phi = \sum_{\xi \in E} \xi \cdot [\phi = \xi] \quad \text{and} \quad \int \phi = \sum_{\xi \in E} \xi \cdot \int [\phi = \xi] \,. \tag{3.3}$$

Consequently $\| \int \phi \|_E \le \|f\|^*$, and $\int : \mathcal{E} \otimes E \to E$ has a unique extension by continuity to a linear map $\int : \mathcal{L}^1_E \to E$.

Exercise 3.19 If $\mathcal{E}$ does not consist of step functions, for instance, if $\int$ is a Radon measure, let $\mathcal{S}$ denote the simple integrable functions. They are the step functions over the ring of integrable sets. Define $\mathcal{S} \otimes E$ naturally in analogy with Equation (3.1), and show that this vector space is dense in $\mathcal{L}^1_E$. The previous exercise provides a unique linear map $\int : \mathcal{L}^1_E \to E$ that is majorized by $\| \ \|^*$. Show that this map agrees on $\mathcal{E} \otimes E$ with the integral defined in Equation (3.2) and provides the desired extension.

In the sequel $\| \ \|^\bullet$ is some order–continuous mean, for instance the Daniell–Stone mean of some order–continuous elementary integral.

Exercise 3.20 Any function in $\mathcal{E}^{\Uparrow}$ is measurable.

Exercise 3.21 Let A be any integrable set. There exists an integrable subset $\underline{A} \subset A$ whose complement in A is negligible with the following property: if ϕ is an elementary integrand such that $\|\phi \cdot A\|^\bullet = 0$ then ϕ vanishes on $\underline{A}$. We call $\underline{A}$ the ***support*** of $\| \ \|^\bullet$ on A.

Exercise 3.22 There exists a collection $\mathcal{C}$ of mutually disjoint integrable sets whose union has locally negligible complement in the sense that $\|A \cdot (\bigcup \mathcal{C})^c\|^\bullet = 0$ for all integrable sets A.

III.4 Measurable Sets

According to our convention of Section I.2:, sets are identified with their indicator functions: *a set is measurable if its indicator function is measurable.*

Exercise 4.1 A set F is measurable if for any integrable set A and $\epsilon > 0$ there are an integrable set $A_0 \subset A$ with $\|A \setminus A_0\|^* < \epsilon$ and a function $\overline{\phi} \in \overline{\mathcal{E}}$ which equals 1 on $A_0 \cap F$ and 0 on $A_0 \setminus F$.

Proposition 4.2 *Let $A_1, A_2, \ldots$ be a countable family of measurable sets. Then A_1^c, $\bigcap_{n=1}^{\infty} A_n$, and $\bigcup_{n=1}^{\infty} A_n$ are measurable as well. In other words, the measurable sets form a σ–algebra.*

Proof.

$$A_1^c = 1 - A_1 ,$$

$$\bigcap_{n=1}^{\infty} A_n = \bigwedge_{n=1}^{\infty} A_n = \lim_{N \to \infty} \bigwedge_{n=1}^{N} A_n ,$$

and

$$\bigcup_{n=1}^{\infty} A_n = \bigvee_{n=1}^{\infty} A_n = \lim_{N \to \infty} \bigvee_{n=1}^{N} A_n ,$$

in the sense that the set on the left has the indicator function on the right, which is measurable by the established in Section 2. ∎

While the definition of measurablity of a function is quite intuitive, its straight reformulation for a set in Exercise 4.1 is not. The next little result provides a very simple criterion.

Proposition 4.3 *A set M is measurable if and only if its intersection with every integrable set A is integrable.*

Proof. If $M \cap A$ is integrable then it is measurable on A. The condition is thus sufficient. Conversely, if M is measurable and A integrable then $M \cap A$ is measurable and has finite mean, so is integrable (3.1). ∎

A finite linear combination of (indicator functions of) measurable sets is clearly measurable. It is a measurable function that takes only finitely many values, every one in a measurable set. Such functions are called ***simple measurable functions***. In other words, the measurable simple functions are the step functions over the σ–algebra of measurable sets.

The simple measurable functions relate to the arbitrary measurable functions in much the same way the simple integrable functions relate to the arbitrary integrable functions (see Proposition II.7.5 on p. 63):

Proposition 4.4 *Let f be a real–valued function.*
(i) If f is measurable then the sets $[f > r]$, $[f \geq r]$, $[f < r]$, and $[f \leq r]$ are measurable for any number r.
(ii) Let $\mathcal{D}$ be a set of reals that is dense in $\mathbb{R}$. If the sets $[f > d]$ are measurable for all $d \in \mathcal{D}$ then f is measurable.
(iii) A measurable function f is the pointwise limit of a sequence (f_n) of measurable simple functions.

Proof. This is but a repetition of the argument of Proposition II.7.5. Note that we use below only the following two properties of the measurable functions: they form a lattice that contains the constants, and they are closed under taking pointwise limits of sequences. For the first claim, observe that

$$\lim_{n\to\infty} 1 \wedge (n(f - f \wedge 1))$$

is measurable, in view of these permanence properties. It is zero at any point x where $f(x) \leq 1$ and 1 at any point x where $f(x) > 1$; in other words, this limit is the (indicator function of the) set $[f > 1]$, which is therefore measurable. The set $[f > r]$ equals $[f/r > 1]$ when $r > 0$ and is thus measurable as well. $[f > 0] = \bigcup_{n=1}^{\infty}[f > 1/n]$ is measurable. Next, $[f \geq r] = \bigcap_{n>1/r}[f > r - 1/n]$, $[f < -r] = [-f > r]$, and $[f \leq -r] = [-f \geq r]$. Finally, when $r \leq 0$ then $[f > r] = 1 - [-f \geq -r]$, etc.
(ii–iii): If $[f > d]$ is measurable for all $d \in \mathcal{D}$ then $[f > r]$ is measurable for all $r \in \mathbb{R}$; simply take a sequence (d_n) in $\mathcal{D}$ that decreases to r and observe that $[f > r] = \bigcup_n [f > d_n]$. The sets $[f \leq r] = [f > r]^c$ etc. are then measurable as well. With this in mind let f_n be the step function over measurable sets

$$f_n = \sum_{k=-2^{2n}}^{2^{2n}} k2^{-n} \cdot [k2^{-n} < f \leq (k+1)2^{-n}]. \qquad (*)$$

The sets $[k2^{-n} < f \leq (k+1)2^{-n}] = [k2^{-n} < f] \cap [(k+1)2^{-n} < f]^c$ are then measurable. Then so are their linear combinations f_n. These evidently converge pointwise to f, so f is measurable. The following is now nearly obvious: ∎

Corollary 4.5 *A function f is measurable if and only if the sets $[f > d]$ are measurable for every dyadic rational d.*

Supplements and Additional Exercises

Exercise 4.6 Show that a set $F \subset \mathbb{R}$ is Lebesgue measurable if and only if

$$\lambda^*(S\backslash F) + \lambda^*(S \cap F) = \lambda^*(S) \qquad (C)$$

for every set $S \in \mathcal{A}$. If so, this equality is satisfied for all integrable subsets S and then for every set $S \subset \mathbb{R}$. Show directly that the sets F that satisfy (C) for all $S \subset \mathbb{R}$ form a σ–algebra.

Exercise 4.7 Let $\mathcal{D}$ be a set of reals dense in $\mathbb{R}$. A function f is measurable if and only if the sets $[f > r]$ are measurable for all $r \in \mathcal{D}$. Then f is measurable if and only if the sets $[f \geq r]$ are measurable for all $r \in \mathbb{R}$. Devise two more criteria along these lines.

Condition (C), though lacking in intuitive appeal, is often used to define the measurablity of sets F, and the statement of Exercise 4.7 to define the measurability of a function f. In other words, a function f is defined to be Lebesgue measurable if

$$\lambda^*(S\backslash[f > d]) + \lambda^*(S \cap [f > d]) = \lambda^*(S)$$

for all subsets S of $\mathbb{R}$ and all $d \in \mathbb{R}$.

Exercise 4.8 Using this definition of measurability show that the sum of two, and the limit of a sequence of, measurable functions is measurable.

Exercise 4.9 Show that $A \subset \mathbb{R}$ is Lebesgue measurable if and only if any or all of its translates $A + a$ are, $a \in \mathbb{R}$ (See Exercise II.7.15).

Convergence in Measure

We do not make use of the results of this subsection later on, so it may be skipped on first reading. For simplicity's sake we assume that the mean is σ–finite.

Definition 4.10 *A sequence (f_n) of functions converges in measure*[4] *to f if, for every $\delta > 0$ and every integrable set A,*

$$\lim_{n\to\infty} \Big\| [f - f_n > \delta] \cap A \Big\|^* = 0 .$$

$\mathcal{L}^0$ denotes the space of a.e. finite measurable functions, equipped with the topology of convergence in measure.

Theorem 4.11 *Let (f_n) be a sequence of measurable functions. (i) If (f_n) converges in mean* **or** *almost everywhere to f then it converges in measure to f.*

[4] A different definition is frequently found: $f_n \to f$ in measure if $\| [|f - f_n| > \delta] \|^* \xrightarrow[n\to\infty]{} 0$ for all $\delta > 0$. It is somewhat easier to state, but also less desirable; it is not true, for instance, that an a.e. convergent sequence converges in measure in this sense.

(ii) Conversely, if (f_n) converges in measure to f then it has a subsequence converging to f almost everywhere.

Proof. If $\|f - f_n\|^* \xrightarrow[n\to\infty]{} 0$ then, since $\delta \cdot [|f - f_n| > \delta] \le |f - f_n|$,

$$\Big\| [|f - f_n| > \delta] \Big\|^* \le \frac{1}{\delta} \cdot \Big\| |f - f_n| \Big\|^* \xrightarrow[n\to\infty]{} 0 ,$$

and hence $f_n \to f$ in measure. Assume now that $f_n \to f$ a.e. and let an integrable set A and $\epsilon > 0$ be given. There is an integrable $A_0 \subset A$ with $\|A - A_0\|^* < \epsilon$ on which (f_n) converges uniformly. If n is so large that $\sup\{|f(x) - f_n(x)| : x \in A_0\} < \delta$ then $\| A \cap [|f - f_n| > \delta] \|^* \le \|A - A_0\|^* < \epsilon$: again (f_n) converges in measure.
Lastly, assume that (f_n) converges in measure to f. Let A_k be a sequence of mutually disjoint integrable sets whose union is the whole space — since the mean is σ–finite, such exists. Find $N_{i,k} \in \mathbb{N}$ so that for $n \ge N_{i,k}$

$$\big\|A_k \cap [|f_n - f| > 2^{-i}]\big\|^* < 2^{-i} ,$$

and set

$$n_j = \bigvee_{i,k \le j} N_{i,k} .$$

Then the set

$$N = \bigcup_k \left(A_k \cap \bigcap_{I\in\mathbb{N}} \bigcup_{i>I} [|f_{n_i} - f| > 2^{-i}] \right)$$

is negligible. For $x \notin N$, say $x \in A_k$, we have

$$x \in \bigcup_{I\in\mathbb{N}} \bigcap_{i>I} [|f_{n_i} - f| \le 2^{-i}] ,$$

which says that for i sufficiently large $|f_{n_i}(x) - f(x)| \le 2^{-i}$. ∎

Corollary 4.12 (The General Dominated Convergence Theorem) *A sequence (f_n) of integrable functions converges in mean if and only if it converges in measure and is uniformly integrable.*

Proof: Exercise. ∎

Exercise 4.13 For every integrable set A (in an adequate cover) and measurable functions f, g set

$$d_A(f, g) = \|A \cdot 1 \wedge |f - g|\|^* .$$

(i) Show that d_A is a pseudometric, and that $f_n \xrightarrow[n\to\infty]{} f$ in measure if and only if $d_A(f_n, f) \to \infty$ for all integrable sets A (in the adequate cover).
(ii) A sequence of measurable functions is ***Cauchy in measure*** if $d_A(f_n, f_m) \xrightarrow[m,n\to\infty]{} 0$

for all integrable sets A (equivalently, for all integrable sets A in an adequate cover). Show that the a.e. finite measurable functions form a vector lattice and algebra that is complete in measure: every sequence Cauchy in measure converges in measure.

Exercise 4.14 Let (A_k) be a sequence of non–negligible integrable sets whose union is the ambient space, and define the pseudometric of convergence in measure by

$$\rho_0(f,g) = \sum_{k=1}^{\infty} \frac{2^{-k}}{\|A_k\|^*} \|A_k|f-g| \wedge 1\|^* \ .$$

Show that $f \mapsto \rho_0(f,0)$ has all the properties of a mean except positive–homogeneity, and that

$$\lim_{r \to 0} \rho_0(r \cdot f, 0) = 0 \quad \forall f \in \mathcal{L}^0 \ .$$

Show that $(f_n) \to f$ in measure if and only if $\rho_0(f_n, f) \xrightarrow[n\to\infty]{} 0$. Show that $(\mathcal{L}^0, \rho_0)$ is a complete pseudometric space in which $\mathcal{E}$ is dense.

III.5 Baire and Borel Functions

The Concept of a Closure under algebraic or order properties is well known to the reader. Let us review it briefly, though, in preparation of the concept below of a closure under limits of convergent sequences.

Suppose, for example, that G is a subset of some vector space. If we want to investigate the linear relations between the elements of G we have to take linear combinations of elements of G, which will take us outside G, in general. So we adjoin to G all linear combinations of elements of G to obtain the ***vector space generated by G***, and inside that we can carry on our investigation.

Next, let G be a collection of real–valued functions on some set and suppose we are interested in the algebraic relations between the functions in G. In order to investigate them we have to take both linear combinations and products of functions in G; but that will lead us outside G, in general. The natural thing to do is to adjoin to G first all finite products of functions in G and then all finite linear combinations of such products. We obtain a ring containing G, in fact the smallest one, which is called the ***ring generated*** or ***spanned by G***. We can now study the algebraic relations of G inside this ring.

It is generally easier, though, to define the ring generated by G as the smallest ring containing G, i.e., as the intersection of all rings containing G. This intersection make sense, because there does exist at least one ring containing G, to wit, the ring of all real–valued functions. If the functions in G are all bounded then so are the functions in the ring generated by G. Instead of producing a bound for every linear combination of products in G it is easier to use this "soft" argument: the collection of all bounded functions is a ring containing G and thus contains the ring generated by G.

Similarly, there is the vector space (algebra, lattice, vector lattice, lattice ring, etc.) generated by G. It is the intersection of all vector spaces (algebras, lattices, vector lattices, lattice rings, etc.) containing G.

Closure under Limits. We can make new measurable functions in the following way. We adjoin to the elementary functions $\mathcal{E}$ the pointwise limits of sequences in $\mathcal{E}$ and get a larger class of functions, all measurable by Egoroff's theorem. We adjoin to the functions in this class the pointwise limits of their sequences and get an even larger class. We continue on — these three words are given a precise meaning in Exercise 5.2. Since we are talking pointwise convergence at every point rather than almost everywhere, the functions we so obtain are all measurable for every single mean on $\mathcal{E}$; the question arises what structure they have in common. Note that we cannot expect to arrive at every function in this way; for instance, we will not get the (indicator function of the) bad set B of Section II.8, since that set is not Lebesgue measurable. We would hope that these functions have some kind of local smoothness in common that can be described without reference to means. Alas, there seems to be no intuitively understandable description of this kind. We simply do as often before: work so much with these functions that we end up thinking we know what they are. Anyway, what counts in the end are their permanence properties, and those we shall exhibit in detail.

To get a better description of the functions above than by the procedure of adjoining more and more limits of more and more sequences, let us introduce the notions of a sequentially closed set of functions and of the sequential closure of a family of functions. A collection $\mathcal{B}$ of functions is ***sequentially closed*** if the limit of any pointwise convergent sequence in $\mathcal{B}$ belongs to $\mathcal{B}$ as well. The intersection of any family $\mathfrak{F}$ of sequentially closed collections of functions is sequentially closed; for if (f_n) is a sequence in $\bigcap \mathfrak{F}$ with pointwise limit f then $f \in \mathcal{F}$ for all $\mathcal{F} \in \mathfrak{F}$ as $\mathcal{F}$ is sequentially closed, and thus $f \in \bigcap \mathfrak{F}$. Given a collection $\mathcal{E}$ of functions there is thus a smallest sequentially closed collection $\mathcal{B}(\mathcal{E}) = \mathcal{E}^\Sigma$ of functions containing $\mathcal{E}$, to wit, the intersection of all sequentially closed collections of functions that contain $\mathcal{E}$. (There is at least one such collection, the collection $\mathbb{R}^S$ of all functions on the ambient set S.) It is reasonable to call $\mathcal{E}^\Sigma$ the ***sequential closure*** or ***sequential span*** of $\mathcal{E}$. Another name for them is the ***$\mathcal{E}$–Baire functions.*** It is clear that it is exactly the $\mathcal{E}$–Baire functions so defined that we were after with the somewhat vague prescription of adjoining more and more limits of more and more sequences to $\mathcal{E}$.

Exercise 5.1* (i) $\mathcal{E} \subset \mathcal{E}' \implies \mathcal{E}^\Sigma \subset \mathcal{E}'^\Sigma)$. (ii) $(\mathcal{E}^\Sigma)^\Sigma = \mathcal{E}^\Sigma$.

Exercise 5.2 $\mathcal{E}^\Sigma$ can be constructed by transfinite induction as follows: Set $\mathcal{E}_0 = \mathcal{E}$. If $\mathcal{E}_\alpha$ has been defined for all ordinals $\alpha < \beta$ define $\mathcal{E}_\beta$ to be the set of all functions that are limits of a sequence in $\mathcal{E}_\alpha$ if β is the successor of α; if β is not a successor define $\mathcal{E}_\beta = \bigcup_{\alpha<\beta} \mathcal{E}_\alpha$. (The functions in $\mathcal{E}_1$ are known as ***$\mathcal{E}$-Baire class 1*** functions, those in $\mathcal{E}_2$ as ***$\mathcal{E}$-Baire class 2*** functions, etc.) Show that $\mathcal{E}_\beta = \mathcal{E}^\Sigma$ if β exceeds the first uncountable ordinal ω_1.

Let us begin the investigation by showing that algebraic and order properties are passed on to the sequential closure.

Proposition 5.3 *Let $\mathcal{E}$ be any family of real–valued functions.*
(i) Any function f in $\mathcal{E}^\Sigma$ is in the sequential closure of a countable subfamily $\mathcal{E}_f \subset \mathcal{E}$. The carrier $[f \neq 0]$ of f can thus be covered by the carriers of countably many functions in $\mathcal{E}$.
(ii) Suppose $\mathcal{E}$ is closed under $+, -, \cdot, \wedge, \vee$, under taking absolute values, **or** *under chopping. Then so is $\mathcal{E}^\Sigma$.*

Proof. (i): The collection $\mathcal{E}'$ of functions f that lie in the sequential closure of a countable family $\mathcal{E}_f \subset \mathcal{E}$ is sequentially closed; for if $\mathcal{E}' \ni f_n \to f$ then f clearly belongs to the sequential closure of the countable family $\bigcup_n \mathcal{E}_{f_n}$. Thus $\mathcal{E}^\Sigma \subset \mathcal{E}'$, which is the first claim. The carrier of f is contained in the countable union $\big\{[\phi \neq 0] : \phi \in \mathcal{E}_f\big\}$, because the functions whose carriers are contained in this set is plainly a sequentially closed family containing $\mathcal{E}_f$.
(ii): Let $*$ stand for anyone of the operations $+, -, \cdot, \vee, \wedge$. Set

$$\mathcal{E}_* = \{f \in \mathcal{E}^\Sigma : f * \phi \in \mathcal{E}^\Sigma \text{ and } \phi * f \in \mathcal{E}^\Sigma \quad \forall \phi \in \mathcal{E}\} .$$

Since $\mathcal{E}$ is closed under $*$, $\mathcal{E}_*$ contains $\mathcal{E}$. Also, $\mathcal{E}_*$ is sequentially closed. For if (f_n) is a sequence in $\mathcal{E}_*$ that converges pointwise to some function f, then $(f_n * \phi)$ converges to $f * \phi$ and $(\phi * f_n)$ converges to $\phi * f$. Since $\phi * f_n, f_n * \phi \in \mathcal{E}^\Sigma \quad \forall n$, the limits $f * \phi, \phi * f$ belong to $\mathcal{E}^\Sigma$ as well. But that means that f is in $\mathcal{E}_*$. Therefore $\mathcal{E}_* = \mathcal{E}^\Sigma$. That is to say, for $f \in \mathcal{E}^\Sigma$ and $\phi \in \mathcal{E}$, $f * \phi, \phi * f \in \mathcal{E}^\Sigma$. In other words, $\mathcal{E}$ is contained in the collection

$$\mathcal{E}_*^* = \{g \in \mathcal{E}^\Sigma : f * g \in \mathcal{E}^\Sigma \text{ and } g * f \in \mathcal{E}^\Sigma \quad \forall f \in \mathcal{E}^\Sigma\} .$$

Now $\mathcal{E}_*^*$ is seen to be sequentially closed in exactly the way $\mathcal{E}_*$ was. Thus it equals $\mathcal{E}^\Sigma$. This again says that $\mathcal{E}^\Sigma$ is closed under $*$.
We leave it to the reader to show that if $\mathcal{E}$ is closed under taking the absolute value or under chopping then so is $\mathcal{E}^\Sigma$. ∎

Corollary 5.4 *If $\mathcal{A}$ is a ring of sets then its sequential closure is a σ–ring, i.e., a ring closed under countable unions — and then automatically under countable intersections.*
If $\mathcal{A}$ is an algebra of sets then its sequential closure is a σ–algebra; the same conclusion holds if $\mathcal{A}$ is a ring and the whole space can be covered by a countable subfamily of $\mathcal{A}$.
An algebra (ring) of sets is a σ–algebra (σ–ring) if and only if it is sequentially closed.

Proof. The collection of all subsets of the ambient set is sequentially closed and contains $\mathcal{A}$, so it contains its sequential closure: $\mathcal{A}^\Sigma$ consists of sets. Since $\mathcal{A}$ is

closed under taking infima (intersections) and suprema (unions), so is $\mathcal{A}^{\Sigma}$. To see that it closed under taking relative complements, apply the previous proof with the operation $A * B = A \setminus B$. In summary, $\mathcal{A}^{\Sigma}$ is a sequentially closed ring, which is easily seen to be a σ–ring.
If $\mathcal{A}$ is an algebra, i.e., a ring containing the constant function 1, then $\mathcal{A}^{\Sigma}$ is a σ–ring containing the constant function 1, i.e., a σ–algebra. $\mathcal{A}^{\Sigma}$ will contain the whole ambient space if and only if the latter can be covered by a countable subfamily of $\mathcal{A}$ (Proposition 5.3, (i)).
It is left to be shown that a σ–algebra or σ–ring $\mathcal{F}$ is sequentially closed. This is plain. Namely, if $\mathcal{F} \ni F_n \to F$ then $F = \bigcap_N \bigcup_{n \geq N} F_n \in \mathcal{F}$. ∎

Exercise 5.5* The sequential closures of $C[\mathbb{R}^n]$, $C_b[\mathbb{R}^n]$, $C_0[\mathbb{R}^n]$, $C_{00}[\mathbb{R}^n]$, $C_{00}^{\infty}[\mathbb{R}^n]$, and the polynomials all coincide. A function in this sequential closure is called a ***Baire function*** on $\mathbb{R}^n$. Of special importance are the Baire functions on the line. The subsets of the line that are Baire functions are called ***Baire sets***.

Corollary 5.6 *Suppose $\mathcal{E}$ is a ring, or a vector lattice closed under chopping.*
(i) Then $\mathcal{E}^{\Sigma}$ is both. Furthermore, $\phi \circ f \in \mathcal{E}^{\Sigma}$ for all $f \in \mathcal{E}^{\Sigma}$ and any Baire function ϕ on the line that vanishes at zero[5].
(ii) The sets in $\mathcal{E}^{\Sigma}$ form a σ–ring $\mathcal{R}(\mathcal{E})$. They form a σ–algebra if and only if there exists a countable subfamily of $\mathcal{E}$ whose pointwise supremum is strictly positive.
(iii) For $f \in \mathcal{E}^{\Sigma}$ the sets $[f > r]$ and $[f < -r]$ belong to $\mathcal{R}(\mathcal{E})$ provided $r \in \mathbb{R}$ is strictly positive. Conversely, if the sets $[f > d]$ and $[f < -d]$ belong to $\mathcal{R}(\mathcal{E})$ for every strictly positive dyadic rational d then $f \in \mathcal{E}^{\Sigma}$.
(iv) Suppose $\mathcal{R}(\mathcal{E})$ is a σ–algebra. Then a function f belongs to $\mathcal{E}^{\Sigma}$ if and only if the set $[f > r]$ belongs to $\mathcal{R}(\mathcal{E})$ for every $r \in \mathbb{R}$.

Proof. (i): The uniform closure $\overline{\mathcal{E}}$ of $\mathcal{E}$ clearly belongs to $\mathcal{E}^{\Sigma}$, and thus $(\overline{\mathcal{E}})^{\Sigma} = \mathcal{E}^{\Sigma}$. By Proposition I.3.10 on p. 11, $\overline{\mathcal{E}}$ is a lattice ring, and then so is $\mathcal{E}^{\Sigma}$. The next claim is true if ϕ is a polynomial with zero constant term, an arbitrary polynomial if $\mathcal{E}$ is σ–finite, for then the constants belong to $\mathcal{E}^{\Sigma}$. The family of functions ϕ such that $\phi \circ f \in \mathcal{E}^{\Sigma}$ for $f \in \mathcal{E}^{\Sigma}$ is sequentially closed, therefore it contains the sequential closure of these polynomials, which is the class of Baire functions or of Baire functions vanishing at 0 as the case may be.
(ii): The (indicator functions of the) sets in $\mathcal{E}^{\Sigma}$ form a sequentially closed family. The claim follows from 5.4. (iii) and (iv) are left as exercises. ∎

Proposition 5.7 *Let $\| \;\|^*$ be a mean on a ring $\mathcal{E}$ of bounded functions.*
(i) Any $(\mathcal{E}, \| \;\|^*)$–integrable function f coincides $\| \;\|^*$–almost everywhere with some $\mathcal{E}$–Baire function.
(ii) Let $\| \;\|^{\sharp}$ be another mean, and assume that $\|\phi\|^* \leq \|\phi\|^{\sharp}$ for all $\phi \in \mathcal{E}_+$.

---

[5] This restriction is superfluous if $\mathcal{E}$ is ***σ–finite***, meaning that there is a countable subfamily of $\mathcal{E}$ whose pointwise supremum is strictly positive.

Then this inequality holds on $\mathcal{E}^{\Sigma}$ *:* $\|f\|^* \le \|f\|^\sharp \quad \forall f \in \mathcal{E}^{\Sigma}$ *. In particular, if* $\|\ \|^*$ *and* $\|\ \|^\sharp$ *agree on* $\mathcal{E}_+$ *then they agree on all* $\mathcal{E}$*–Baire functions.*

Proof. (i): There is a sequence (ϕ_n) in $\mathcal{E}$ that converges to f almost everywhere. Set $h = \limsup_n \phi_n$. This function agrees with f almost everywhere. It is not quite an $\mathcal{E}$–Baire function, though, inasmuch as it might take the values $\pm\infty$. We have to change the definition a little. The set C of points where (ϕ_n) converges,

$$\bigcap_{k\in\mathbb{N}} \bigcup_{N\in\mathbb{N}} \bigcap_{m,n>N} [|\phi_n - \phi_m| \le 1/k] ,$$

is a $\mathcal{E}$–Baire function, and $h = \limsup_n C \cdot \phi_n$ meets the description.
(ii): Let f be an $\|\ \|^\sharp$–integrable function. If $\phi_n \in \mathcal{E}$ converges to f in $\|\ \|^\sharp$–mean then it converges in $\|\ \|^*$–mean as well, and f is $\|\ \|^*$–integrable. Thus $\|f\|^* = \lim \|\phi_n\|^* = \lim \||\phi_n|\|^* \le \lim \||\phi_n|\|^\sharp = \|f\|^\sharp$.
Next let $h \in \mathcal{E}_+^{\Sigma}$. There is a countable family $\{\phi_1, \phi_2, \ldots\} \subset \mathcal{E}$ whose carriers cover the carrier of h. The $\mathcal{E}$–Baire functions

$$h_n = h \wedge n \cdot \bigcup_{1\le\nu\le n} [|\phi_\nu| > 1/n]$$

are integrable for any mean on $\mathcal{E}$ (Theorem II.5.5), and so $\|h_n\|^* \le \|h_n\|^\sharp \quad \forall n$. The h_n increase pointwise to h. If $\|h\|^\sharp < r$ then $\sup_n \|h_n\|^* \le \sup_n \|h_n\|^\sharp < \infty$, and by the Monotone Convergence Theorem (h_n) converges to h both in $\|\ \|^*$–mean and in $\|\ \|^\sharp$–mean. Hence $\|h\|^* < r$. We conclude that $\|h\|^* \le \|h\|^\sharp$. The inequality follows for all $\mathcal{E}$–Baire functions h, since the solidity of the means implies $\|h\|^* = \big\||h|\big\|^* \le \big\||h|\big\|^\sharp = \|h\|^\sharp$. ∎

Measurability on a Σ–algebra

The last statement (iv) of Corollary 5.6 describes the membership of a function in a sequentially closed *algebra* $\mathcal{E}$ in terms of its relation to the σ–algebra of sets in $\mathcal{E}$. It repeats the measurability criterion (ii) of Proposition 4.4 on p. 82 and gives rise to the following

Definition 5.8 *Let* $\mathcal{F}$ *be a* σ*–algebra on the ambient space* S*. A function* $f : S \to \mathbb{R}$ *is said to be* ***measurable on*** $\mathcal{F}$*, or* $\mathcal{F}$***–measurable*** *if* $[f > r] \in \mathcal{F}$ *for all* $r \in \mathbb{R}$*. The collection of all* $\mathcal{F}$*–measurable functions is denoted by* $\mathcal{M}(\mathcal{F})$*. Continuing the use of the word, we call an* $\mathcal{F}$*–measurable function* ***simple*** *if it takes only finitely many values (each one of them evidently in a set of* $\mathcal{F}$*).*

Warning *We have now two notions of measurability:* ***for*** *a mean* $(\mathcal{E}, \|\ \|^*)$ *and* ***on*** *a* σ*–algebra* $\mathcal{F}$*. We shall carefully both distinguish and compare them.*

Exercise 5.9* Let $\mathcal{F}$ be a σ–algebra, f a function, and $\mathcal{D}$ a dense set of reals. The following are equivalent: (i) $[f > r] \in \mathcal{F} \quad \forall\, r \in \mathbb{R}$; (ii) $[f \le r] \in \mathcal{F} \quad \forall\, r \in \mathbb{R}$; (iii) $[f > r] \in \mathcal{F} \quad \forall\, r \in \mathbb{R}$; (iv) $[f \ge r] \in \mathcal{F} \quad \forall\, r \in \mathbb{R}$; (v) $[f > d] \in \mathcal{F} \quad \forall\, d \in \mathcal{D}$; (vi) $[f \le d] \in \mathcal{F} \quad \forall\, d \in \mathcal{D}$; (vii) $[f > d] \in \mathcal{F} \quad \forall\, d \in \mathcal{D}$; (viii) $[f \ge d] \in \mathcal{F} \quad \forall\, d \in \mathcal{D}$; (ix) $[d_1 < f \le d_2] \in \mathcal{F} \quad \forall\, d_1, d_2 \in \mathcal{D}$; (x) f is $\mathcal{F}$–measurable; (xi) $f^{-1}(B) \in \mathcal{F}$ for all Baire sets $B \subset \mathbb{R}$; (xii) $\phi \circ f$ is $\mathcal{F}$–measurable for all Baire functions ϕ on the line.

$\mathcal{M}(\mathcal{F})$ is sequentially closed: If $f_n \in \mathcal{M}(\mathcal{F})$ converges pointwise to f then $[f > r] = \bigcap_N \bigcup_{n \ge N} [f_n > r]$, which is evidently a set in $\mathcal{F}$, and so f is measurable on $\mathcal{F}$. We know from Exercise II.2.17 on p. 40 that the simple $\mathcal{F}$–measurable functions form a lattice algebra of bounded functions. Its sequential closure consists of $\mathcal{F}$–measurable functions. In fact, due to Corollary 5.6 (iv) **the class of real–valued $\mathcal{F}$–measurable functions is precisely the sequential closure of the $\mathcal{F}$–measurable simple functions**. The $\mathcal{F}$–measurable functions are therefore closed under addition, multiplication, taking infima, etc.

Exercise 5.10 Show this directly, using only Definition 5.8.

Generating σ–algebras. Σ–algebras are most frequently given by specifying a generator: the ***σ–algebra generated*** by a collection $\mathcal{C}$ of functions is the smallest σ–algebra on which every function in $\mathcal{C}$ is measurable. Such exists: it is the intersection of all σ–algebras on which every function in $\mathcal{C}$ is measurable, and $\mathcal{C}$ is called its ***generator***.

For instance, the ***Borel σ-algebra*** $\mathcal{B}^\bullet[S]$ of a topological space S is defined as the σ–algebra generated by the topology, in other words as the smallest σ–algebra containing every open set. Its elements are the ***Borel sets***, and the functions measurable on it are called ***Borel measurable*** or ***Borel functions***. There is also the σ–algebra $\mathcal{B}^*[S]$ generated by the continuous bounded functions. It is called the ***Baire σ–algebra***. The functions measurable on it are called ***Baire measurable*** or ***Baire functions***. Every topological space is considered to be naturally equipped with these two σ–algebras.

Exercise 5.11* (i) $\mathcal{B}^*[S] \subset \mathcal{B}^\bullet[S]$. On a separable metric space S the two σ–algebras coincide. If the two σ–algebras agree then every σ–additive measure on $C_b[S]$ is order–continuous, and so is every mean on $C_b[S]$. [Hint: $C_b^{\uparrow}[S] = C_b^{\Uparrow}[S]$.] (ii) A function is Baire measurable if and only if it belongs to the sequential closure of the continuous bounded functions.

Exercise 5.12 Let S be a locally compact space[6]. A function h on S is ***lower semicontinuous*** if the sets $[h > r]$ are open for all $r \in \mathbb{R}$. The set $\mathcal{H}$ of lower semicontinuous functions is closed under addition, finite infima and suprema, and under multiplication with positive reals. It coincides with $C_{00}^{\Uparrow}[S]$ (See page 47). The collection $\mathcal{H}_b - \mathcal{H}_b$ of differences of bounded lower semicontinuous functions is a vector lattice closed under chopping. The Borel–measurable functions are the sequential closure of it. If $\|\ \|^\bullet$ is an order–continuous mean on $C_{00}[S]$ then any Borel function is $\|\ \|^\bullet$–measurable.

---

[6] Hausdorff and normal would suffice for this exercise.

Exercise 5.13 Let $\mathcal{E}$ be a ring of bounded functions on some set S. Give S the topology generated by the sets $[\phi > r]$, $\phi \in \mathcal{E}, r \in \mathbb{R}$. Then $\mathcal{E}^{\Uparrow}$ consists of lower semicontinuous functions, and the sequential closure of $\mathcal{E}_b^{\Uparrow} - \mathcal{E}_b^{\Uparrow}$ is exactly the class of Borel measurable functions. They are measurable for any order–continuous mean on $\mathcal{E}$. Since for a continuous function f the set $[f > r]$ is open and thus

Example 5.14 Prepare uncountably many copies of the unit interval, say $I_\alpha, \alpha \leq \aleph_1$, and let K be their product. This is a compact space. The polynomials in finitely many coordinates form an algebra separating the points, so they are uniformly dense in $C[K]$. Every continuous function on K therefore depends on countably many coordinates at most, and then so does every Baire function (Proposition 5.3, (i)). The set of points in K that have at least one coordinate in $(0, 1/2)$ is an open subset, therefore Borel. As it depends on all coordinates it is not Baire.

Exercise 5.15 Let f be a measurable function with values in a complete uniform space E and ϕ a Baire function on E. Show that $\phi \circ f$ is measurable.

One the line the Baire and Borel sets coincide. An open interval (a, b) is the pointwise limit of the continuous functions ϕ_n that equal zero off (a, b) and one on $[a+1/n, b-1/n]$ and are linear on $[a, a+1/n]$ and $[b-1/n, b]$. (a, b) is thus a Baire set, and then then so is every open set on the grounds that it can be written as a countable union of open intervals.
An open interval, and then every open set, is, by a similar argument, an $\mathcal{E}[\mathbb{R}]$–Baire function. The σ–algebra generated by the step functions is thus again the Borel σ–algebra. The sequential closure of the polynomials contains $C[\mathbb{R}]$ (Exercise I.3.15 on p. 14 (iii)) and thus all Borel functions. We see that the polynomials, the continuous functions, and the step functions all have the same sequential closure, namely the Borel measurable functions.

Let $\mathcal{M}$ be an algebra of functions that is sequentially closed, for instance the class of functions measurable on some σ–algebra, or the $(\mathcal{E}, \| \;\|^*)$–measurable functions of some mean $\| \;\|^*$. For any Baire function ψ on the line and any $f \in \mathcal{M}$, the composition $\psi \circ f$ belongs to $\mathcal{M}$. Indeed, the collection of functions ϕ with $\phi \circ \mathcal{M} \subset \mathcal{M}$ is plainly sequentially closed and contains the polynomials, so it contains all Borel functions. In other words,

Proposition 5.16 *A sequentially closed algebra of functions is closed under (left–) composition with Baire functions.*

As a consequence note the following. If f is an $(\mathcal{E}, \| \;\|^*)$–measurable or $\mathcal{F}$–measurable function and B a Borel subset of the line then the set $[f \in B]$ is measurable. Indeed, this set is nothing but the composition $B \circ f = 1_B \circ f$ of a Borel function with f.

Exercise 5.17 (The Monotone Class Theorem) Let $\mathcal{M}$ be a vector space of real–valued functions that is closed under pointwise limits of increasing or decreasing bounded sequences and contains the constants, a ***monotone class***.
If $\mathcal{I} \subset \mathcal{M}$ is a collection of bounded functions that is closed under taking finite products then $\mathcal{M}$ contains the σ–algebra generated by $\mathcal{I}$ and every bounded function measurable on it.

Example 5.18 Suppose $(\mathcal{E}, \| \ \|^*)$ is a mean. We have seen in Proposition 5.7 above that a $(\mathcal{E}, \| \ \|^*)$–integrable function agrees almost everywhere with an $\mathcal{E}$–Baire function. One might hope that an $(\mathcal{E}, \| \ \|^*)$–measurable function agrees almost everywhere with an $\mathcal{E}$–Baire measurable function, i.e., a function measurable on the σ–algebra generated by $\mathcal{E}$. This is true if $\mathcal{E}^\Sigma$ contains the constants, but not in general. To see this let S be an uncountable set, $\mathcal{E}$ the lattice ring of functions with finite carrier, and $\| \ \|^*$ the Daniell mean for counting measure (see Example II.2.22 on p. 41). Only the empty set is $\| \ \|^*$–negigible, and every function is $\| \ \|^*$–measurable. Not every function is measurable on the σ–algebra $\mathcal{F}$ generated by $\mathcal{E}$. Indeed, $\mathcal{F}$ is the collection of all sets of the form $A \cup B$ where A is countable and B has countable complement, on the ground that the sets of this description form a σ–algebra on which the functions of $\mathcal{E}$ are measurable, evidently the smallest one. Not every set belongs to $\mathcal{F}$.

Exercise 5.19* On the positive side, assume that the ambient set S is σ–finite for the set $\mathcal{E}$ of elementary integrands. That is to say, assume that there is a countable subfamily of $\mathcal{E}$ whose pointwise supremum is strictly positive. Then $1 \in \mathcal{E}^\Sigma$, and a function is measurable on the σ–algebra $\mathcal{E}^\sigma$ generated by $\mathcal{E}$ if and only if it belongs to the sequential closure $\mathcal{E}^\Sigma$ of $\mathcal{E}$.
Let now $\| \ \|^*$ be a mean on $\mathcal{E}$. The σ–algebra of $(\mathcal{E}, \| \ \|^*)$–measurable sets is the σ–algebra generated by $\mathcal{E}$ and the $\| \ \|^*$–negligible sets. A set A belongs to it if and only if there is a set $A' \in \mathcal{E}^\sigma$ with $\| |A - A'| \|^* = 0$. A function is $(\mathcal{E}, \| \ \|^*)$–measurable if and only if it differs $\| \ \|^*$–negligibly from a function in $\mathcal{E}^\Sigma$.

Exercise 5.20 Let $\mathcal{E}$ be a lattice ring of bounded functions and $\mathcal{F}$ the σ–algebra generated by $\mathcal{E}$. A set F belongs to $\mathcal{F}$ if and only if it has the form $F = A \cup B$ with $A \in \mathcal{E}^\Sigma$ and $B^c \in \mathcal{E}^\Sigma$.

Measurable Spaces and Maps. A ***measurable space*** is a pair $(S, \mathcal{F})$ consisting of a set S and a σ–algebra $\mathcal{F}$ of subsets of S. Let $(T, \mathcal{G})$ be a second measurable space. A map $\Phi : S \to T$ is called $\mathcal{F}/\mathcal{G}$–measurable if the preimage of every set in $\mathcal{G}$ belongs to $\mathcal{F}$: $f^{-1}(\mathcal{G}) \subset \mathcal{F}$. This should remind the reader of the definition of continuity of a map between topological spaces.

Exercise 5.21 Let $\Phi : (S, \mathcal{F}) \to (T, \mathcal{G})$ be measurable. For every $\mathcal{G}$–measurable function f, $f \circ \Phi$ is $\mathcal{F}$–measurable.

Exercise 5.22 Let A be a Lebesgue measurable set. Show that there exist Borel sets $\overline{A}$ and $\underline{A}$ with $\underline{A} \subset A \subset \overline{A}$ and Lebesgue negligible difference $\overline{A} \setminus \underline{A}$.

III.6 Further Properties of Daniell's Mean

Let us summarize the development so far. Daniell's construction produces the seminorm $\| \ \|^*$. Since $\| \ \|^*$ is a mean, the Dominated Convergence Theorem holds on the mean closure $\mathcal{L}^1$ of the elementary functions, which implies that $\mathcal{L}^1$ is rich. Since the elementary integral is majorized by $\| \ \|^*$ it has a unique extension by continuity to all of $\mathcal{L}^1$.

It is not hard to see that any extension of the elementary integral to a vector lattice of functions on which the Monotone and Dominated Convergence Theorems hold is an extension by continuity with respect to some mean (Exercise II.6.9). The question arises whether we did the best possible. Are there not, perhaps, other means on $\mathcal{E}$ that majorize the elementary integral, smaller than the Daniell mean if possible so they would produce even more integrable functions? Indeed there are. In fact, the Daniell mean is the *largest* mean that agrees with the elementary integral on $\mathcal{E}_+$. However, the other means one can think of are not *canonical* in any sense, while Daniell's mean is, *by virtue of being maximal.* These thoughts would lead the thorough thinker into investigating means *per se*. We won't indulge in this — nobody has so far — but make a few observations about them.

Maximality. Let $\| \ \|^*$ be a mean for a ring $\mathcal{E}$ of bounded functions. There exists a maximal mean that agrees with $\| \ \|^*$ on $\mathcal{E}_+$. Namely, if $\mathfrak{M}$ is the collection of all means that coincide with $\| \ \|^*$ on $\mathcal{E}_+$, set

$$\|f\|^\sharp = \sup\left\{\|f\|^\flat : \ \| \ \|^\flat \in \mathfrak{M}\right\}.$$

Let us show that $\| \ \|^\sharp$ is countably subadditive: if $\sum_n \|f_n\|^\sharp < r < \infty$ then, for every single $\| \ \|^\flat \in \mathfrak{M}$, $\sum_n \|f_n\|^\flat < r$ and $\|\sum_n f_n\|^\flat < r$; consequently $\|\sum_n f_n\|^\sharp < r$. The other properties of a mean refer only to its behaviour on $\mathcal{E}$ and thus are evident. $\| \ \|^\sharp$ is evidently the largest mean that agrees with $\| \ \|^*$ on $\mathcal{E}_+$. Let us call $\| \ \|^*$ a ***maximal mean*** if it coincides with $\| \ \|^\sharp$. The means $\| \ \|^*$ we shall encounter in the sequel will be maximal — and if one isn't we simply replace it by its associated maximal mean $\| \ \|^\sharp$. We know from Proposition 5.7 on p. 88 (ii) that $\| \ \|^*$ and $\| \ \|^\sharp$ agree on the class $\mathcal{E}^\Sigma$ of $\mathcal{E}$–Baire functions.

Lemma 6.1 *For all numerical functions f on the ambient space*

$$\|f\|^\sharp = \inf\left\{\|h\|^* : \ |f| \le h \in \mathcal{E}^\Sigma\right\}. \tag{6.1}$$

Proof. Let $\|f\|^\uparrow$ denote the right–hand side of (6.1). It defines a mean. Again everything is obvious except possibly the countable subadditivity. To show it, let $f_n \ge 0$ be functions with $\sum_n \|f_n\|^\uparrow < r < \infty$. There are functions $h_n \in \mathcal{E}^\Sigma$ with $f_n \le h_n$ and $\sum_n \|h_n\|^* < r$. The function $h = \sum_n h_n$ belongs to $\mathcal{E}^\Sigma$, exceeds $\sum_n f_n$, and has $\|h\|^* < r$. We conclude that $\|\sum_n f_n\|^\uparrow < r$ and that $\| \ \|^\uparrow$ is countably subadditive.
Clearly $\| \ \|^\uparrow \le \| \ \|^\sharp$. For the converse inequality assume $\|f\|^\uparrow < r$. There exists $|f| \le h \in \mathcal{E}^\Sigma$ with $\|h\|^* < r$. Then $\|f\|^\sharp \le \|h\|^\sharp = \|h\|^* < r$ and consequently $\|f\|^\sharp \le \|f\|^\uparrow$. ∎

Proposition 6.2 *(i) The Daniell mean of an elementary integral is maximal. It can thus be characterized as the unique maximal mean that agrees with the elementary integral on $\mathcal{E}_+$.*

Let now $\| \ \|^*$ be a maximal mean for a ring $\mathcal{E}$ of bounded functions.
(ii) For any positive function f with finite mean there exists an integrable function $\overline{f} \geq f$ with $\|\overline{f}\|^* = \|f\|^*$. Every such $\overline{f}$ is called an integrable ***upper envelope*** *of f. It can be chosen in $\mathcal{E}^\Sigma$, and if f is a set it can be chosen to be a set as well.*
*(iii) For every integrable function f there exist $\mathcal{E}$–Baire functions $\underline{f}, \overline{f}$ with $\underline{f} \leq f \leq \overline{f}$ and $\|\overline{f} - \underline{f}\|^* = 0$. They are called an $\mathcal{E}$–**Baire upper and lower envelope**, respectively. If f is a set, $\underline{f}$ and $\overline{f}$ can be chosen to be sets as well.*
(iv) The mean is ***continuous along arbitrary increasing sequences*** *in the following sense: for every increasing sequence (f_n) of positive numerical functions with pointwise supremum f*

$$\sup_n \|f_n\|^* = \|f\|^* .$$

Proof. (i): The very construction of the Daniell mean shows that it is maximal, in view of Equation (6.1):

$$\|f\|^* = \inf\left\{\|h\|^* : |f| \leq h \in \mathcal{E}^\uparrow\right\} \geq \inf\left\{\|h\|^* : |f| \leq h \in \mathcal{E}^\Sigma\right\} = \|f\|^{*\sharp} .$$

(ii): According to (6.1) there are $f \leq h_n \in \mathcal{E}_+^\Sigma$ with $\|h_n\| \leq \|f\| + 1/n$. Each of the h_n is integrable (Exercise II.7.7), and so is their pointwise infimum $\overline{f}$, which evidently has the same mean as f and belongs to $\mathcal{E}^\Sigma$. If f is a set then replacing $\overline{f}$ by $[\overline{f} \geq 1] \in \mathcal{E}^\Sigma$ produces a smaller $\mathcal{E}$–Baire envelope.
(iii): Let h be an $\mathcal{E}$–Baire function that coincides almost everywhere with f, as provided by Proposition 5.7 on p. 88, (i). There is another $\mathcal{E}$–Baire function h_0 with $|f - h| \leq h_0$ and $\|h_0\|^* = 0$. The functions $\underline{f} = h - h_0$ and $\overline{f} = h + h_0$ evidently meet the description. If f is a set then $[\underline{f} > 0]$ and $[\overline{f} \geq 1]$ are idempotent lower and upper $\mathcal{E}$–Baire envelopes, respectively.
(iv): If $\sup_n \|f_n\|^* = \infty$ there is nothing to prove. In the remaining case we employ the upper the envelopes $\overline{f}_n$ of (ii). Replacing if necessary $\overline{f}_n$ by $\bigwedge_{\nu \geq n} \overline{f}_\nu$, we see that the $\overline{f}_n$ can be chosen to increase with n. Their pointwise limit g is integrable by the Monotone Convergence Theorem, since $\sup_n \|\overline{f}_n\|^* = \sup_n \|f_n\|^* < \infty$. Since $f \leq g$, $\|f\|^* \leq \|g\|^* = \sup_n \|f_n\|^*$. The reverse inequality is evident from the solidity of $\| \ \|^*$. ∎

Exercise 6.3 If $0 \leq f_n \uparrow f$ merely a.e., where all the f_n are a.e. defined then again $\|f_n\|^* \uparrow \|f\|^*$.

Exercise 6.4 Is there a minimal mean that agrees with a given mean on $\mathcal{E}_+$?

Exercise 6.5 Define the notion of a maximal order–continuous mean and show that the Daniell–Stone mean $\| \ \|^\bullet$ of Equation (II.3.3) on p. 48 is one. Show that a maximal order–continuous mean is again continuous along arbitrary increasing sequences.

Strict Monotonicity. A mean $\| \ \|^*$ on $\mathcal{E}$ is called ***strictly increasing*** if for any two $\| \ \|^*$–integrable functions f, g

$$\left\{ \begin{array}{c} 0 \le f \le g \text{ a.e.} \\ \text{and} \\ \|f\|^* = \|g\|^* \end{array} \right\} \quad \text{imply} \quad f = g \quad \text{a.e.}$$

In other words, if $0 \le f \le g$ and $f < g$ on some non–negligible set then $\|f\|^* < \|g\|^*$. Any mean is ϵ–close to a strictly increasing mean with the same integrable functions (Exercise IV.4.10 on p. 118). That Daniell's mean has this property is a nearly trivial observation, but it will come in handy later on.

Proposition 6.6 *Daniell's mean* $\| \ \|^*$ *for an elementary integral* $\int$ *is strictly increasing.*

Proof. Let f, g be two integrable functions with $0 \le f \le g$ a.e. By Theorem II.6.1 on p. 58, (ii), $\|f\|^* = \|g\|^*$ implies $0 = \|g\|^* - \|f\|^* = \int^* g - \int^* f = \int g - \int f = \int (g - f) = \int^* |g - f| = \|g - f\|^*$, which says that $g - f$ is negligible. ∎

Exercise 6.7 Let $\| \ \|^*$ be a maximal and strictly increasing mean. Two integrable upper envelopes of a function $0 \le f \in \mathcal{F}^*[\| \ \|^*]$ agree almost everywhere.

Exercise 6.8 Let $S = \{1, \ldots, n\}$ and $\mathcal{E}$ the collection of all bounded functions on S; i.e., $\mathcal{E} = \mathbb{R}^n$. Then $\| \ \|_u$ is a mean on $\mathcal{E}$ that is not strictly increasing.

III.7 The Procedures of Lebesgue and Carathédory

Daniell's method of extending a measure or elementary integral is not the only one. In fact, there are several, none of them canonical. This section explains two of the more common ones. This gives the reader a chance to connect with the literature and, so we hope, will convince her that no method produces any better results than any other. The section contains no new facts, so it may be omitted on first reading.

We will discuss only Lebesgue measure λ on the ring $\mathcal{A} = \mathcal{A}[\mathbb{R}]$ of finite unions of intervals, and leave the generalization to arbitrary measures to the reader. As hinted at in Section I.1, Lebesgue's idea towards extending the integral to a larger class of integrable functions was to start by measuring more sets than merely intervals and their finite unions, then to assign an integral to a step function ϕ over the enlarged class of sets by the time–honored formula

$$\int \phi(x) = \sum \textit{height–of–step} \times \textit{measure–of–step} \,, \tag{7.1}$$

and lastly to integrate even more functions by applying the Riemann squeeze or some similar device. Carathéodory later generalized Lebesgue's ideas to abstract measures. His is still the approach taken in most textbooks on integration, notably in Halmos' influential book [6] and its descendants.

Let us start by analyzing (7.1). Suppose we have succeeded in extending the measure of intervals to a set function λ on a large class $\mathcal{R}$ of sets. If we require as little from the integral as its linearity then the step functions over $\mathcal{R}$ must form a vector space; else we cannot even formulate linearity. This in turn forces $\mathcal{R}$ to be a ring and λ to be additive on it (see Exercise II.2.17 on p. 40). Thus if we want a linear integral, and if we want to find it by first measuring sets and then applying (7.1), then the pair $(\mathcal{R}, \lambda)$ must be a positive measure in the sense of Proposition II.2.13 on p. 39. How can we construct such a pair $(\mathcal{R}, \lambda)$?

Just as the main body of Chapter II rested on the "Little Limit Lemma" II.1.1 for the elementary integral, does progress come from the observation that $(\mathcal{A}, \lambda)$ is a σ–additive positive measure in the sense of Proposition II.2.16:

Lemma 7.1 *For every increasing sequence (A_n) in $\mathcal{A}$ whose union A belongs to $\mathcal{A}$*

$$\lambda(A) = \sup_n \lambda(A_n) .$$

Inasmuch as A_n, A are step functions on the line, this is but a special case of Lemma II.1.1, (i) — still, it is a good exercise to give a direct proof.

Let us now define the *outer measure* λ^* on the collection $\mathcal{A}^\uparrow$ of sets H that are countable unions of sets in $\mathcal{A} = \mathcal{A}[\mathbb{R}]$ by

$$\lambda^*(H) = \sup\{\lambda(A) : H \supset A \in \mathcal{A}\} \quad , \quad H \in \mathcal{A}^\uparrow .$$

Just as the σ–additivity of the elementary integral resulted in the additivity of $\int^*$ on $\mathcal{E}^\uparrow$ does Lemma 7.1 yield the additivity of λ^* on $\mathcal{A}^\uparrow$. It is too early to rejoice, though, because $\mathcal{A}^\uparrow$ is not a ring. So we extend the definition of λ^* to all sets $F \subset \mathbb{R}$, which do form a ring, by

$$\lambda^*(F) = \inf\{\lambda^*(H) : F \subset H \in \mathcal{A}^\uparrow\} .$$

This is, of course nothing but the "up–and–down–procedure" of Daniell, restricted to sets. The trouble is that this extension is not additive anymore, it is merely subadditive. What to do? The following idea succeeds: let us replace the "up–and–down–procedure" by a "down–and–up–procedure." The resulting *inner measure* λ_* will be superadditive, or so we hope. The sets F with $\lambda_*(F) = \lambda^*(F)$ will form a ring, on which both λ_* and λ^* are additive. If this ring is large, progress has been made.

Let us flesh out the last paragraph with a few details. $\mathcal{A}_\downarrow$ denoting the collection of sets that are countable intersections of elementary sets, we define the Lebesgue inner measure λ_* first on sets $K \in \mathcal{A}_\downarrow$ by

$$\lambda_*(K) = \inf\{\lambda(A) : K \subset A \in \mathcal{A}\}$$

and then on arbitrary sets $F \subset \mathbb{R}$ by

$$\lambda_*(F) = \sup\{\lambda_*(K) : F \supset K \in \mathcal{A}_\downarrow\}.$$

The properties of λ^*, λ_* are analogue to those of $\int^*, \int_*$ established in Section II.3:

Proposition 7.2 *(i) $\mathcal{A}^\uparrow$ is closed under taking finite intersections and countable unions, and λ^* is continuous along increasing sequences on $\mathcal{A}^\uparrow$ and thus is additive on $\mathcal{A}^\uparrow$;*

(ii) λ^* is countably subadditive: $\quad \lambda^*(\bigcup_{n=1}^\infty F_n) \le \sum_{n=1}^\infty (F_n)$;

(iii) λ^* is increasing along arbitrary increasing sequences: $F_n \uparrow F$ implies $\lambda^*(F_n) \uparrow \lambda^*(F)$;

(iv) λ_* is superadditive and continuous along arbitrary decreasing sequences of sets. Furthermore,

$$\lambda_*(F) \le \lambda^*(F) \quad \forall\, F \subset \mathbb{R}, \quad \textit{and} \quad \lambda_*(A) = \lambda(A) = \lambda^*(A) \quad \forall\, A \in \mathcal{A}\,.$$

Proof. The proofs of Proposition II.3.2 and Exercise II.3.10 apply in particular to idempotent functions, i.e., sets. It is a good exercise to redo these proofs, judiciously replacing every function by an appropriate set, thus arriving at the claims above. Another and faster way to establish the claims (i)–(iv) is to notice that

$$\lambda^*(F) = \int^* F \text{ and } \lambda_*(F) = \int_* F \quad \forall\, F \subset \mathbb{R}\,. \tag{$\binom{*}{*}$}$$

The statements (i)—(v) then appear as restrictions of known facts about functions to sets. Let us prove $\binom{*}{*}$. First $\lambda^*(H) = \int^* 1_H$ for all $H \in \mathcal{A}^\uparrow$. Indeed, let $(A_n), (A'_n)$ be sequences of subsets of H in $\mathcal{A}$ such that (a) $\bigcup A_n = H$ and (a') $\sup \lambda(A'_n) = \lambda^*(H)$. Then the sequence of sets $A''_n = \bigcup_{\nu=1}^n (A_\nu \cup A'_\nu)$ has both properties (a) and (a'). Since $\mathcal{A}^\uparrow \subset \mathcal{E}^\uparrow$, we get from Lemma II.3.1

$$\int^* H = \lim \int A''_n \, d\lambda = \lim \lambda(A''_n) = \lambda^*(H)\,.$$

Next, $\lambda^*(F) \ge \int^* F$, since the infimum in the definition of $\lambda^*(F)$ is extended over the subset $\mathcal{A}^\uparrow$ of $\mathcal{E}^\uparrow$, which latter appears in the definition of $\int^* F$. Only the reverse inequality $\lambda^*(F) \le \int^* F$ needs to be shown. Now if $\int^* F < r$ then there is an $h \in \mathcal{E}^\uparrow$ with $\int^* h < r$. For any $\epsilon > 0$, the set

$$[h > 1 - \epsilon] \le \frac{1}{1-\epsilon} \cdot h$$

belongs to $\mathcal{A}^\uparrow$ and contains F. Consequently,

$$\lambda^*(F) \le \lambda^*([h > 1-\epsilon]) = \int^* [h > 1-\epsilon] \le \frac{1}{1-\epsilon} \int^* h < \frac{r}{1-\epsilon} .$$

Since $\epsilon > 0$ was arbitrary, the desired inequality results. The second equality of $\binom{*}{*}$ is shown similarly. ▮

Corollary 7.3 *For every $F \subset \mathbb{R}$, $\lambda^*(F) = \int^* F$ and $\lambda_*(F) = \int_* F$.*

Lebesgue–Carathéodory would, of course, prove Proposition 7.2 directly from the definitions of λ^*, λ_*. Once their properties are established it is clear what our sought–after ring $\mathcal{R}$ should be:

Proposition 7.4 *The collection $\mathcal{R}$ of sets $F \subset \mathbb{R}$ with*

$$\lambda_*(F) = \lambda^*(F) < \infty \tag{L}$$

forms a δ–ring of sets, and the restriction λ of λ^* (or of λ_*) to it is σ–additive.

This is evident. By Exercise II.6.2 and Corollary 7.3, $\mathcal{R}$ consists precisely of the integrable sets in our sense, which by Proposition II.7.1 on p. 61 form a δ–ring. Some treatments of measure theory go this route and adopt (L) as the definition of integrability of a set F.

There is a slight aesthetical flaw in this definition: one has to develop the properties of both λ^* and λ_*. At the corresponding juncture in Daniell's treatment the simple observation that

$$\int_* f = -\int^* (-f)$$

made it easy to deduce the properties of $\int_*$ from those of $\int^*$ (Exercise II.3.14). This is not possible here, since we are limited to sets: if F is a set, $-F$ is not one. Carathéodory avoids the use of λ_* by the following consideration. Suppose we have succeeded in restricting λ^* to a large ring $\mathcal{R}$ of sets on which it is additive. For any set $F \in \mathcal{R}$ we have then

$$\lambda^*(S \setminus F) + \lambda^*(S \cap F) = \lambda^*(S) \quad \text{for all} \quad S \in \mathcal{R} . \tag{7.2}$$

We cannot, of course, use this equation as the definition of the integrability of F, since we do not know which sets S belong to $\mathcal{R}$; such a definition would be begging the question. So Carathéodory defines a set F to be integrable if $\lambda^*(F) < \infty$ and

$$\lambda^*(S \setminus F) + \lambda^*(S \cap F) = \lambda^*(S) \quad \text{for } all \text{ subsets } S \subset \mathbb{R} . \tag{C}$$

This "Carathéodory Cut Condition" is *a priori* much more restrictive than (7.2), and it is not even obvious that sets F from $\mathcal{A}$ satisfy it. Yet this definition succeeds:

Exercise 7.5 Show directly that the sets F satisfying (C) form a δ–ring containing $\mathcal{A}$, and that the restriction of λ^* to it is σ–additive.

It is time to show that all roads lead to the same point.

Theorem 7.6 *Let F be a subset of $\mathbb{R}$. The following are equivalent.*
(i) F is integrable in the sense of Section II.7;
(ii) $\lambda_*(F) = \lambda^*(F) < \infty$;
(iii) $\lambda^*(F) < \infty$ and $\lambda^*(S \setminus F) + \lambda^*(S \cap F) = \lambda^*(S)$ for all subsets $S \subset \mathbb{R}$;
(iv) $\lambda^*(F) < \infty$ and $\lambda^*(A \setminus F) + \lambda^*(A \cap F) = \lambda(A)$ for all $A \in \mathcal{A}$;
(v) For every $\epsilon > 0$ there exists a set $A \in \mathcal{A}$ with $\lambda^*(F \bigtriangleup A) < \epsilon$.

Proof. The equivalence of (i) and (ii) is immediate from Corollary 7.3 and Exercise II.6.2.
(ii) $\Longrightarrow$ (iii): If $\lambda^*(S) = \infty$ there is nothing to prove. In the other case let $\overline{S}$ denote the integrable outer envelope of S provided by Exercise II.6.6 or Proposition 6.2 on p. 93. Then

$$\lambda^*(S) = \lambda(\overline{S}) = \lambda(\overline{S}\backslash F) + \lambda(\overline{S} \cap F) \geq \lambda^*(S\backslash F) + \lambda^*(S \cap F) \geq \lambda^*(S)$$

so that equality obtains throughout. Clearly (iii) implies (iv).

(iv) $\Longrightarrow$ (v) $\Longrightarrow$ (i): Let $\epsilon > 0$ be given. There exists a set $H \in \mathcal{A}^\uparrow$ with $F \subset H$ and $\lambda^*(H) < \lambda^*(F) + \epsilon$. H is integrable (Theorem II.5.4) and can be written as the countable union of disjoint sets $A_n \in \mathcal{A}$. Addition of the equalities

$$\lambda^*(A_n \setminus F) + \lambda^*(A_n \cap F) = \lambda(A_n)$$

results in the inequality

$$\lambda^*(H \setminus F) + \lambda^*(H \cap F) \leq \lambda^*(H) \leq \lambda^*(F) + \epsilon\,.$$

Now $H \cap F = F$, and $\|H - F\|^* = \lambda^*(H \setminus F) \leq \epsilon$ follows. For sufficiently large index N and $A = \bigcup_{n=1}^N A_n$ we have $\|H - A\|^* = \lambda^*(H - A) < \epsilon$, which gives $\|F - A\|^* = \lambda^*(F \bigtriangleup A) \leq 2\epsilon$. ∎

We have succeeded with the first part of the program, measuring sets. Let us denote by $\mathcal{A}[\lambda]$ the collection of Lebesgue integrable sets, identified by the equivalent conditions of Theorem 7.6. This is the ring $\mathcal{R}$ we were looking for. $\mathcal{A}[\lambda]$ contains the elementary sets $\mathcal{A}$ but is strictly bigger, because it is a δ–ring, which $\mathcal{A}$ is not. Progress has been made.

Integrating functions is the main purpose of measure theory. The first step is to integrate the step functions over $\mathcal{A}[\lambda]$ — let us again call them the *integrable simple functions* and denote their collection by $\mathcal{E}[\lambda]$. The integral of an integrable simple function is, of course, the sum *step–size* $\times$ *height–of–step*. We know from Exercises II.2.17 or 7.5 that $\mathcal{A}[\lambda]$ is a ring and from Proposition II.2.13 that there is a unique linear map

$$s \mapsto \int s\, d\lambda \ :\ \mathcal{E}[\lambda] \to \mathbb{R}$$

extending λ. It is σ–additive (Proposition II.2.16 on p. 40). The pair $(\mathcal{E}[\lambda], \int\, d\lambda)$ is a positive σ–additive elementary integral; but progress has been made, since $\mathcal{E}[\lambda]$ is much larger than the original lattice ring $\mathcal{E}[\mathbb{R}]$.

We want to integrate more than simple functions. We might think of applying the Riemann squeeze with $\mathcal{E}[\lambda]$ replacing $\mathcal{E}[\mathbb{R}]$; however, this is easily seen to lead to too restriced a class of integrable functions: they would all be bounded above and below by a simple function. Not even positive improperly Riemann integrable functions would be subsumed.

What to do?

The easiest thing is to use Daniell's method on the elementary integral $(\mathcal{E}[\lambda], \int\ )$, even at this juncture. However, we are discussing here the situation where his scheme had not yet been thought of. Lebesgue and Carathéodory had ingenuity at their disposal, we use hindsight: to be integrable, a function must not be too big and must be measurable (Theorem 3.1 on p. 75). What should "measurable" mean? Since at this juncture only the integrable sets are known, we should perhaps define first when a *set* is measurable. Carathéodory simply declares a set F to be λ^*–measurable if it satisfies his Cut Condition (C) of page 99.

Exercise 7.7* A set F satisfies (C) if and only if its intersection with every integrable set is integrable. The collection of such sets forms a σ–algebra.

He then declares a function f to be measurable if the sets $[f > r]$ are measurable for every $r \in \mathbb{R}$ and proves the permanence properties of measurability from this. It is clear from Proposition 4.3 and Corollary 4.5 that Carathéodory's class of measurable functions agrees with ours. Finally, he declares a positive *measurable* function f to be integrable if the supremum of the integrals of simple integrable functions majorized by f is finite, and sets $\int f\, d\lambda$ equal to this supremum. An arbitrary measurable function is declared integrable if both its positive and negative parts f_+, f_- are, and the integral is then defined as $\int f = \int f_+ - \int f_-$. By Exercise 3.8 on p. 78, Carathéodory arrives precisely at the integrable functions of Daniell. The reader can find the details in the books [6] and [9], for example.

Exercise 7.8 Show that the Daniell extension procedure applied to the pair $(\mathcal{E}[\lambda], \int)$ leads no further than to $\mathcal{L}^1$.

Project 7.9 *If μ is a positive σ–additive measure on some ring $\mathcal{R}$ of sets, not necessarily Lebesgue measure on $\mathcal{A}[\mathbb{R}]$, we can clearly define the associated outer and inner measures μ^*, μ_* in the same way. Redo the present section in this context.*

Chapter IV

The Classical Banach Spaces

Historical Note and Motivation. Let us return to the beginning of Chapter II. Given an integrable function f on $(-\pi, \pi]$, we can define its *Fourier coefficients* a_n, b_n by Equation (II.2), and ask in which sense the formal partial sums

$$f_n = \sum_{\nu=0}^{n} a_\nu \cos(\nu x) + b_\nu \sin(\nu x)$$

converge to the original function f. One might hope that $f_n \to f$ at least almost everywhere. It was long known that this is true when f is piecewise continuously differentiable. In 1876 DuBois Reymond [5] showed that one cannot hope for convergence everywhere when f is continuous. Thirty years later Kolmogoroff [7] provided an integrable function whose Fourier series diverges at every single point. This looked very discouraging, until in 1966 Carleson [4] showed that a condition on the size of f will produce almost everywhere convergence. Namely, if one requires that not only f but also its square f^2 is integrable, then $f_n \to f$ a.e. It is now known that it suffices to require that $|f|^p$ be integrable for some $p > 1$.

From the fact that forty years passed between the last two results it is plain that the proofs of Carleson's and subsequent theorems are too complicated to be included in these notes. Let us take them as the motivation to study the spaces $\mathcal{L}^p$ of functions whose p^{th} power is integrable. They are known as the *Classical Banach Spaces* $\mathcal{L}^p$. For the precise definition see Section 3 on p. 108. The $\mathcal{L}^p$–spaces were actually discovered as early as 1906 by F. Riesz [8], who was at the time investigating compact subsets of $\mathcal{L}^1$. They are in many ways superior to $\mathcal{L}^1$. As Banach spaces they are, for instance, reflexive when $1 < p < \infty$, which $\mathcal{L}^1$ is not. $\mathcal{L}^1$ is not in general closed under multiplication, as we know (Exercise II.5.12); Hölder's inequality (Theorem 1.2) permits the estimation in $\mathcal{L}^1$ of products of functions in terms of the $\mathcal{L}^p$–sizes of the factors. There are amazing interpolation results linking the various $\mathcal{L}^p$ and operators between them (Section V.4).

Throughout we consider a fixed mean $\| \; \|^*$ on the lattice ring $\mathcal{E}$. It simplifies things to assume that $\| \; \|^*$ is ***continuous along arbitrary increasing sequences***, and we shall do so. This will for instance be the case if the mean is maximal (see Proposition III.6.2 (iv)). Whether $\| \; \|^*$ is the Daniell mean of some σ–additive elementary integral does not matter for a while — yet it may be helpful to visualize it as that.

K. Bichteler, *Integration – A Functional Approach*, Modern Birkhäuser Classics,
DOI 10.1007/978-3-0348-0055-6_4, © Birkhäuser Verlag 1998

IV.1 The p–norms

The space $\mathcal{L}^p$ can be defined as the space of measurable functions f such that $|f|^p$ is integrable, or as the closure of the elementary integrands under the mean $f \mapsto \left(\left\||f|^p\right\|^*\right)^{1/p}$. By Theorem 2.5 below, this amounts to the same thing. We choose the latter route, because the properties of single p–integrable functions as well as the properties of their collection $\mathcal{L}^p$ are easier to establish this way — in fact they have already been established in the previous chapters when we were developing the properties of an arbitrary mean.

Definition 1.1 *Let f be an almost everywhere defined numerical function.*
(i) For $1 \le p < \infty$, the ***p–norm*** *or* ***p–mean*** *of f is the number*

$$\|f\|_p \overset{\text{def}}{=} \left(\left\||f|^p\right\|^*\right)^{1/p} \quad \in \overline{\mathbb{R}}_+ \; ;$$

$$\|f\|_\infty \overset{\text{def}}{=} \inf\{r \in \overline{\mathbb{R}} : [|f| > r] \quad \textit{is negligible}\}$$

covers the case $p = \infty$ and is called the ***infinity norm*** *or* ***essential supremum norm*** *of f* [1].
(ii) A sequence (f_n) of numerical almost everywhere defined functions with finite p–norm is said to ***converge in p–norm*** *to f (and also* ***in p–mean*** *if $p < \infty$) if*

$$\|f - f_n\|_p \xrightarrow[n\to\infty]{} 0 \, .$$

Note that $\| \ \|_1 = \| \ \|^*$. First a result relating the p–norms to each other:

Theorem 1.2 (Hölder's Inequality) *Let $p, p' \in [1, \infty]$ with $\dfrac{1}{p} + \dfrac{1}{p'} = 1$. (Such a pair of exponents is called a* ***conjugate pair****). Then for any two functions f, g,*

$$\|f \cdot g\|_1 \le \|f\|_p \cdot \|g\|_{p'} \, .$$

Proof. Let us dispense with a few easy cases first. If f or g is negligible then $\left\||fg|\right\|^* = 0$, and the inequality is satisfied; if $p = 1$ and consequently $p' = \infty$ it follows from the inequality

$$|fg| \le |f| \cdot \|g\|_\infty \qquad \text{a.e.}$$

[1] The infimum of the empty set is by convention $+\infty$.

upon application of $\| \;\|^*$. We need to deal thus only with the case $p \neq 1 \neq p'$ and $\|f\|_p \neq 0 \neq \|g\|_{p'}$. To this end consider the inequality

$$a^\lambda \cdot b^{1-\lambda} \leq \lambda{\cdot}a + (1-\lambda)b \,.$$

It holds for any $a, b \in [0, \infty]$ and $\lambda \in [0,1]$, and is left to the reader to establish. We employ it with $a = |f(x)|^p / \|f\|_p^p$, $b = |g(x)|^{p'} / \|g\|_{p'}^{p'}$, and $\lambda = 1/p$, $1 - \lambda = 1/p'$, and obtain

$$\frac{1}{\|f\|_p \|g\|_{p'}} |fg| = \left(\frac{|f|^p}{\|f\|_p^p}\right)^{1/p} \cdot \left(\frac{|g|^{p'}}{\|g\|_{p'}^{p'}}\right)^{1/p'} \leq \frac{1}{p} \cdot \frac{|f|^p}{\|f\|_p^p} + \frac{1}{p'} \cdot \frac{|g|^{p'}}{\|g\|_{p'}^{p'}} \,.$$

We apply $\| \;\|^*$ to this and obtain the desired inequality in the form

$$\frac{1}{\|f\|_p \|g\|_{p'}} \cdot \big\||fg|\big\|^* \leq \frac{1}{p} + \frac{1}{p'} = 1 \,.$$

▮

Exercise 1.3* Let $1 \leq p \leq \infty$, and $1/p + 1/p' = 1$. Then, for any f,

$$\|f\|_p = \sup\{\|fg\|_1 : \|g\|_{p'} \leq 1\} \,.$$

If f is positive and/or measurable then the g in $\{\cdots\}$ can be restricted to be positive and/or measurable.

Exercise 1.4 For $p, q, r \in [1, \infty]$ with $\frac{1}{r} = \frac{1}{p} + \frac{1}{q}$,

$$\|fg\|_r \leq \|f\|_p \cdot \|g\|_q \quad \text{and} \quad \|f\|_p = \sup\{\|f{\cdot}g\|_r : \|g\|_q \leq 1\} \,.$$

To justify the name "p–norm" it must be shown that $\| \;\|_p$ is subadditive:

Theorem 1.5 (Minkowski's Inequality) *For Minkowski's inequalityany two functions f, g*

$$\|f + g\|_p \leq \|f\|_p + \|g\|_p \,, \qquad 1 \leq p \leq \infty \,.$$

Proof. First the case $p = \infty$. If $\|f\|_\infty + \|g\|_\infty < b$ then there are $r, s \in \mathbb{R}$ with sum $< b$ such that both $[|f| > r]$ and $[|g| > s]$ are negligible. Then $[|f + g| > b]$ is negligible and thus $\|f + g\|_\infty < b$.

Now to the case $1 \leq p < \infty$. Let p' be the conjugate exponent of p, defined by $1/p + 1/p' = 1$, or $p' = p/(p-1)$. By Hölder's inequality.

$$\big\||f| \cdot |f + g|^{p-1}\big\|^* \leq \|f\|_p \cdot \||f + g|^{p-1}\|_{p'}$$

and

$$\big\||g| \cdot |f + g|^{p-1}\big\|^* \leq \|g\|_p \cdot \big\||f + g|^{p-1}\big\|_{p'} \,.$$

Upon addition:
$$\begin{aligned}\|f+g\|_p^p &= \big\|(|f+g|)\cdot|f+g|^{p-1}\big\|^* \\ &\le \big\|(|f|+|g|)\cdot|f+g|^{p-1}\big\|^* \\ &\le (\|f\|_p+\|g\|_p)\cdot\big\||f+g|^{p-1}\big\|_{p'} \\ &= (\|f\|_p+\|g\|_p)\cdot\big\|f+g\big\|_p^{p-1},\end{aligned}$$

because $p/p' = p-1$ and $(p-1)p' = p$. We divide both sides by $q = (\|f+g\|_p)^{p-1}$ and arrive at Minkowski's inequality. For this to make sense, q must not equal zero or ∞. Well, if it is zero then there is nothing to prove in the first place. Neither is there if $\|f\|_p$ or $\|g\|_p$ is infinite; but if both are finite then so is q, since $|f+g|^p \le 2^p(|f|^p+|g|^p)$ and thus $\|f+g\|_p < \infty$. ∎

The time has come to list all the properties of $\|\ \|_p$.

Theorem 1.6 *Let $1 \le p \le \infty$. (i) The number $\|f\|_p \in [0,\infty]$ is defined for every numerical function f, and $\|\ \|_p$ is absolute–homogeneous, countably subadditive, solid, and continuous along arbitrary increasing sequences:*

$$0 \le f_n \uparrow f \implies \|f_n\|_p \uparrow \|f\|_p\,. \tag{S}$$

Moreover, $\|\ \|_p$ is finite on elementary functions.
(ii) For $1 \le p < \infty$, $\|\ \|_p$ is a mean, maximal or strictly increasing if $\|\ \|^*$ is.

Proof. It is plain that $\|\ \|_p$ is subadditive, absolute–homogeneous, and solid. The continuity along increasing sequences is easy: if (f_n) is an increasing sequence of positive functions with supremum f then (f_n^p) increases to f^p. Since $\|\ \|^*$ is assumed to be continuous along arbitrary increasing sequences $\|f_n^p\|^* \uparrow \|f^p\|^*$, and (S) follows upon taking p^{th} roots. This takes care of the case $1 \le p < \infty$. For $p = \infty$ we argue as follows: If $0 \le f_n \uparrow f$ and $\|f_n\|_\infty < b \ \ \forall\, n$ then $[f > b] \subset \bigcup [f_n > b]$ is negligible and thus $\|f\|_\infty < b$. Hence $\|\ \|_p$ is continuous along increasing sequences in all cases.
The countable subadditivity follows easily from this and the subadditivity. Indeed, for any sequence of positive numerical functions f_n

$$\Big\|\sum_{n=1}^{\infty} f_n\Big\|_p = \lim_{N\to\infty}\Big\|\sum_{n=1}^{N} f_n\Big\|_p \le \lim_{N\to\infty}\sum_{n=1}^{N}\|f_n\|_p = \sum_{n=1}^{\infty}\|f_n\|_p\,.$$

(ii): To show that $\|\ \|_p$ is a mean for $1 \le p < \infty$, property (M) of page 46 must be established. That is to say, for every sequence (ϕ_n) of positive elementary functions

$$\sup_N \Big\|\sum_{n=1}^{N}\phi_n\Big\|_p < \infty \quad \text{implies} \quad \|\phi_n\|_p \xrightarrow[n\to\infty]{} 0\,. \tag{M}$$

To see this, let $\psi_N = \sum_{n=1}^N \phi_n$. Then $\psi_N^p \in \mathcal{E}^\Sigma$ is $\| \ \|^*$–measurable and majorized by $\|\psi_N\|_u^{p-1} \cdot \psi_N$ and so is $\| \ \|^*$–integrable. The sequence (ψ_N^p) increases and $\sup_N \|\psi_N^p\|^*$ is finite. This sequence is therefore Cauchy with respect to $\| \ \|^*$. In particular,

$$\|\psi_N^p - \psi_{N-1}^p\|^* \xrightarrow[N\to\infty]{} 0 .$$

Since $\phi_N^p \le \psi_N^p - \psi_{N-1}^p$, $\|\phi_N^p\|^* = \|\phi_N\|_p^p \xrightarrow[N\to\infty]{} 0$, and (M) follows. [Note that (M) has no chance of being true if $p = \infty$: the sequence $([n, n+1))$ of elementary functions is a counterexample.]

To see that $\| \ \|_p$ is maximal we check that it equals its associated maximal mean $\| \ \|_p^\sharp$ of Equation (III.6.1) on p. 93. Let $f \ge 0$ have $\|f\|_p < r$, i.e., $\|f^p\|^* < r^p$. There exists $h \in \mathcal{E}_+^\Sigma$ with $f^p \le h$ and $\|h\|^* < r^p$. By Corollary III.5.6 on p. 88, $h^{1/p}$ belongs to $\mathcal{E}_+^\Sigma$. Now $f \le h^{1/p}$ and $\|h^{1/p}\|_p < r$: we conclude that $\|f\|_p^\sharp \le \|f\|_p$. Since the converse inequality is obvious, equality obtains.

The strict monotonicity is best proved after integrability with respect to $\| \ \|_p$ has been discussed and is left as an exercise. ∎

Remark 1.7 Nowhere did we make use of the assumption that $\| \ \|^*$ be additive on $\mathcal{E}_+$. We can apply the construction of $\| \ \|_p$ from $\| \ \|^*$ to any mean $\| \ \|$ that is continuous along increasing sequences, in particular to $\| \ \|_p$—this does not lead anywhere, though.

IV.2 The $\mathcal{L}^p$–spaces

So $\| \ \|_p$ is a solid and countably subadditive seminorm if $1 \le p \le \infty$, and a mean on $\mathcal{E}$ if $1 \le p < \infty$. We can and will talk about $\| \ \|_p$–negligible functions and sets, about the class $\mathcal{F}^p$ of functions with finite p–norm, and about functions defined $\| \ \|_p$–almost everywhere.

It is evident from the definition that a function or set is $\| \ \|_p$–negligible if and only if it is $\| \ \|^*$–negligible. The notion of almost everywhere convergence, in particular, is the same for $\| \ \|^*$ and for any of the $\| \ \|_p$. The word function means "almost everywhere defined function" as before.

Since only the solidity and the countable subadditivity of $\| \ \|^*$ were used in the proof of Theorem II.4.6, we get it for free in the present case:

Proposition 2.1 *Let $1 \le p \le \infty$. $\mathcal{F}^p$ is a vector lattice closed under chopping and $\| \ \|_p$ is a solid and countably subadditive seminorm on it. The pair $(\mathcal{F}^p, \| \ \|_p)$ is complete. Moreover, any p–norm convergent sequence (f_n) in $\mathcal{F}^p$ has a subsequence that converges almost everywhere to a norm–limit.*

We concentrate now for a while on the case of finite p.

Definition 2.2 *Let $1 \le p < \infty$. A function f is p–integrable if it is the limit in p–mean of a sequence of elementary functions. The collection of p–integrable functions is denoted by $\mathcal{L}^p$, and by $\mathcal{L}^p[\mathcal{E}, \| \ \|^*]$ or $\mathcal{L}^p[\lambda] = \mathcal{L}^p[\mathbb{R}]$ if we want to exhibit the ingredients.*

In other words, $\mathcal{L}^p$ is the $\| \ \|_p$–mean closure of $\mathcal{E}$, just as $\mathcal{L}^1$ was the $\| \ \|^*$–mean closure of $\mathcal{E}$.

The proofs of Section II.5, Chapter II made use only of the fact that $\| \ \|^*$ was a mean on $\mathcal{E}$. So is $\| \ \|_p$. Therefore the permanence properties of $\mathcal{L}^p$ are for free:

Theorem 2.3 *Let f, f' be p–integrable functions and $r \in \mathbb{R}$. Then $f + f'$, rf, $f \vee f'$, $f \wedge f'$, $|f|$ and $f \wedge 1$ are p–integrable; so is $f \cdot f'$ if f or f' is bounded. $\mathcal{L}^p$ is complete in p–mean; moreover, every p–mean convergent sequence has an almost everywhere convergent subsequence. The Monotone and Dominated Convergence Theorems hold: If (f_n) is a sequence of p–integrable functions that converges pointwise almost everywhere to some function f then f is p–integrable and (f_n) converges to f in p–mean, provided (f_n) is increasing or decreasing and* $\sup_n \|f_n\|_p < \infty$ **or** *there is a function $g \in \mathcal{L}^p$ with $|f_n| \le g \ \forall n \in \mathbb{N}$* **or** *$\{f_n\}$ is uniformly $\| \ \|_p$–integrable.*

Exercise 2.4 (f_n) converges to f in p–mean if and only if it converges to f in measure and $\{f_n\}$ is uniformly p–integrable.

What have the spaces $\mathcal{L}^p$ to do with each other?

Theorem 2.5 *Let $1 \le p, q < \infty$. A function f is p–integrable if and only if the function $f \cdot |f|^{(p/q)-1}$ is q–integrable. In particular, f is p–integrable if and only if $f \cdot |f|^{p-1}$ is integrable.*

Proof. Assume f is p–integrable. There is a sequence (ϕ_n) of elementary functions with $\sum \|\phi_n\|_p < \infty$ and $f = \sum \phi_n$ a.e. (Exercise II.5.8). Set

$$\psi_n = \left(\sum_{k=1}^{n} \phi_n \right) \cdot \left| \sum_{k=1}^{n} \phi_n \right|^{(p/q)-1} .$$

This is an $\mathcal{E}$–Baire function. Clearly $\psi_n \xrightarrow[n\to\infty]{} f$ almost everywhere, and (ψ_n) is dominated by the function $h = (\sum |\phi_n|)^{p/q}$, which has q–mean

$$\left\| (\sum |\phi_n|)^{p/q} \right\|_q = \left(\left\| \sum |\phi_n| \right\|_p \right)^{p/q} \le \left(\sum \|\phi_n\|_p \right)^{p/q} < \infty .$$

By Exercise II.5.14 ψ_n is q–integrable, and by the Dominated Convergence Theorem so is f.
The reverse implication follows from interchanging p with q. ∎

Corollary 2.6 *For $1 \le p < \infty$, the p–integrable and the integrable sets coincide.*

Let us now talk about measurability. It is plain what the definition should be: for $1 \le p < \infty$, when $\| \;\|_p$ is a mean, a function f should be declared to be measurable in p–mean if, for every (p–mean–) integrable set A and $\epsilon > 0$, there is a (p–mean–) integrable subset A_0 of A with $\|A - A_0\|_p < \epsilon$ on which f is the uniform limit of elementary functions. Since two means on $\mathcal{E}$ with the same integrable sets have the same measurable sets (Proposition III.4.3) and measurable functions (Corollary III.4.5), though, there is no novelty in this definition: measurability is the same for all the p–means. The results of Chapter III did use only the structure of $\mathcal{E}$ and the fact that $\| \;\|^*$ was a mean. They carry over. For instance,

Theorem 2.7 *Let $1 \le p < \infty$. A function f is p–integrable if and only if it is measurable and has finite p–mean and σ–finite carrier.*

Exercise 2.8 A measurable function f belongs to $\mathcal{L}^p$ iff $|f|^p$ belongs to $\mathcal{L}^1$.

We turn to the case $p = \infty$. The closure of $\mathcal{E}$ under the essential supremum norm $\| \;\|_\infty$ is too small to be useful. In its stead consider the space $\mathcal{L}^\infty$ of all *measurable* functions having finite essential supremum norm; its natural seminorm is $\| \;\|_\infty$. This is a space with interesting properties:

Exercise 2.9* Show that $\mathcal{L}^\infty$ is an algebra and that $\| \;\|_\infty$ is submultiplicative, defining this notion. In fact, $\mathcal{L}^\infty$ is a complete seminormed lattice algebra.

Exercise 2.10* Let $1 \le p \le \infty$. If $f \in \mathcal{L}^p$ and $g \in \mathcal{L}^\infty$ then $f \cdot g \in \mathcal{L}^p$ and $\|f \cdot g\|_p \le \|f\|_p \cdot \|g\|_\infty$.

Exercise 2.11 Exercise 2.10 can be read as saying that every $g \in \mathcal{L}^\infty$ acts by pointwise multiplication as a continuous linear operator on $\mathcal{L}^p$, whose operator norm is less than $\|g\|_\infty$. Show that the operator norm of multiplication with $g \in \mathcal{L}^\infty$ actually equals $\|g\|_\infty$.

Additional Exercises

Exercise 2.12 Describe the closure of $\mathcal{E}$ in the essential supremum norm.

Exercise 2.13 Let A be a set with $\|A\|^* = 1$, and let f be a function that vanishes off A. Then $\|f\|_\infty = \lim_{p \to \infty} \|f\|_p$.

Exercise 2.14 Go through the exercises of Sections II.4–II.7 and determine which of the results you proved there obtain for the p–means $(1 \le p < \infty)$.

Exercise 2.15 C_{00}^∞ is dense in $\mathcal{L}^p[\lambda] = \mathcal{L}^p[\mathbb{R}]$ for $1 \le p < \infty$. $\mathcal{L}^p[\mathbb{R}]$ is separable if $1 \le p < \infty$, while $\mathcal{L}^\infty[\mathbb{R}]$ is not. $\mathcal{L}^p[\mathbb{R}]$ is translation invariant.

Exercise 2.16 $\mathcal{L}^\infty[\| \;\|^*]$ is separable if and only if it is finite dimensional.

IV.3 The L^p–spaces

It is by now evident that two almost everywhere defined functions that differ only negligibly are the same for all practical purposes. So let us identify them: for any measurable almost everywhere defined function f we denote by $\dot{f}$ the *class* of all functions that agree almost everywhere with f, i.e.,

$$\dot{f} = \{f' : f' = f \text{ a.e.}\} .$$

The sum and scalar product of classes are defined in the obvious way: $\dot{f} + \dot{g}$ is the class of $f + g$, and $r \cdot \dot{f}$ is the class of $r \cdot f$. These classes do not depend on the representatives $f \in \dot{f}, g \in \dot{g}$ chosen; for instance, if $f = f'$ almost everywhere and $g = g'$ almost everywhere then the class of $f + g$ is clearly the same as the class of $f' + g'$. We can define the p–norm of a class as

$$\|\dot{f}\|_p = \|f\|_p ,$$

where $f \in \dot{f}$; again, this number will not depend on the choice of $f \in \dot{f}$ (Proposition II.4.3 on p. 49). Let us then denote by

$$\mathrm{L}^p = \{\dot{f} : f \in \mathcal{L}^p\} , \qquad 1 \le p \le \infty$$

the collection of classes of almost everywhere defined functions in $\mathcal{L}^p$. In the language of Exercise I.7.6, L^p is the quotient of $\mathcal{L}^p$ by its subspace of negligible functions, and $\|\ \|_p$ is the quotient norm. It is left to the reader to establish

Proposition 3.1 $(\mathrm{L}^p, \|\ \|_p)$ *is a Banach space for* $1 \le p \le \infty$.

There is an additional structure on $\mathcal{L}^p$, the order. The question arises whether it is inherited by the set L^p of classes. It is. Let us say that the class $\dot{g}$ exceeds, or is bigger than, the class $\dot{f}$ if a representative of $\dot{g}$ exceeds a representative of $\dot{f}$ almost everywhere. In other words,

$$\dot{f} \le \dot{g} \iff f \le g \text{ a.e.}$$

for any — and then clearly all — representatives $f \in \dot{f}$, $g \in \dot{g}$. This order makes L^p an order complete Banach lattice in the sense of Definition 3.2 below.

Before showing that let us agree on some language concerning order. Let $\mathcal{V}$ be a set. An ***order*** on $\mathcal{V}$ is a reflexive ($v \le v$), antisymmetric ($v \le v'$ and $v' \le v$ imply $v = v'$), and transitive ($u \le v$ and $v \le w$ imply $u \le w$) relation. It is not implied that any two elements v, v' can be compared with this order. "v' exceeds v," "v' is bigger than v," "v is less than v'," "v is smaller than v'," and "$v' \ge v$" all mean "$v \le v'$." A subset $\mathcal{F}$ of an ordered set $\mathcal{V}$ is called ***order bounded*** from above (below) if there is an element $b \in \mathcal{V}$ bigger (smaller) than every member of $\mathcal{F}$. In this case b is called an upper (lower) ***order bound*** for $\mathcal{F}$. An upper (lower) order bound of $\mathcal{F} \subset \mathcal{V}$ that is less (bigger) than all other upper (lower) order bounds, is called a ***least upper bound*** (***greatest lower bound***) and is denoted by $\bigvee \mathcal{F}$ ($\bigwedge \mathcal{F}$). It is evident from the antisymmetry of the order that a least upper bound (greatest lower bound), should it exist, is unique. A least upper bound (greatest lower bound) of the two–point set $\{v, w\} \subset \mathcal{V}$ is called a ***maximum*** (***minimum***) of v and w and is denoted by $v \vee w$ ($v \wedge w$). If any two–point subset of $\mathcal{F} \subset \mathcal{V}$ has an upper (lower) order bound inside $\mathcal{F}$ then $\mathcal{F}$ is ***increasingly directed*** (***decreasingly directed***). If any two–point set in $\mathcal{V}$ — and then any finite set — has both a least upper bound and a greatest lower bound, then the ordered set $(\mathcal{V}, \leq)$ is called a ***lattice***. It is ***order–complete*** if every non–void subset $\mathcal{S}$ that is order bounded above has least upper bound $\bigvee \mathcal{S}$ and every non–void subset $\mathcal{S}$ that is order bounded below has a greatest lower bound $\bigwedge \mathcal{S}$.

What have linear structure, order, and norm to do with each other?

Definition 3.2 *Suppose $\mathcal{V}$ is a real vector space.*
(i) The order $\leq$ on $\mathcal{V}$ is said to be ***compatible with the linear structure****, and the pair $(\mathcal{V}, \leq)$ is called an* ***ordered vector space****, if*

$$v \leq v' \implies v + w \leq v' + w \text{ and } rv \leq rv'$$

for all $v, v', w \in \mathcal{V}$ and $r \in \mathbb{R}_+$.

The ordered vector space $(\mathcal{V}, \leq)$ is called a ***vector lattice*** *or* ***Riesz space*** *if it is a lattice in its order, that is to say, if any two elements have a maximum and a minimum. In this case $v \vee -v$ is denoted by $|v|$ or $¦v¦$ and called the* ***absolute value*** *or* ***variation*** *of v. A vector lattice $(\mathcal{V}, \leq)$ is called* ***order complete*** *if it is order complete in its order.*
(ii) Suppose now that $\| \ \|$ is a seminorm on the ordered vector space $(\mathcal{V}, \leq)$. If

$$-v' \leq v \leq v' \text{ implies } \|v\| \leq \|v'\| \qquad \forall\, v, v' \in \mathcal{V}$$

then the seminorm is called ***solid*** *and the triple $(\mathcal{V}, \leq, \| \ \|)$ is called a* ***seminormed ordered vector space****. If $\mathcal{V}$ is a lattice under the order $\leq$ then the seminorm is solid if and only if $|v| \leq |v'| \implies \|v\| \leq \|v\|' \ \ \forall\, v, v' \in \mathcal{V}$, and $(\mathcal{V}, \leq, \| \ \|)$ is a* ***seminormed vector lattice****.*
(iii) If $(\mathcal{V}, \leq, \| \ \|)$ is a normed vector lattice and (topologically!) complete under the norm then it is called a ***Banach lattice****.*

Exercise 3.3* Let $(\mathcal{V}, \leq)$ be an ordered vector space and $v, v', w \in \mathcal{V}$. Then v, v' have a least upper bound (greatest lower bound) if and only if $v + w, v' + w$ do, and in that case $(v + w) \vee (v' + w) = v \vee v' + w$ ($(v + w) \wedge (v' + w) = v \wedge v' + w$. Next let $\emptyset \neq \mathcal{F} \subset \mathcal{V}$ and $w \in \mathcal{V}$. Then $\mathcal{F}$ has a least upper bound (greatest lower bound) if and only if $\mathcal{F} + w = \{v + w : v \in \mathcal{F}\}$ does, and in that case $\bigvee(\mathcal{F} + w) = \bigvee \mathcal{F} + w$ ($\bigwedge(\mathcal{F} + w) = \bigwedge \mathcal{F} + w$). $\emptyset \neq \mathcal{F} \subset \mathcal{V}$ has a least upper bound if and only if $-\mathcal{F} = \{-v : v \in \mathcal{F}\}$ has a greatest lower bound, and in that case $\bigvee \mathcal{F} = -\bigwedge -\mathcal{F}$. Conclude that a vector lattice is order complete if and only if every non–void set bounded from above has a least upper bound.

Exercise 3.4* Let $(\mathcal{V}, \leq)$ be an ordered vector space. Show that the following are equivalent: (i) $\mathcal{V}$ is a vector lattice; (ii) any two elements $v, w \in \mathcal{V}$ have a least upper bound $v \vee w$ — in that case $-(-v \vee -w)$ is their greatest lower bound $v \wedge w$; (iii) any two elements $v, w \in \mathcal{V}$ have a greatest lower bound $v \wedge w$ — in that case $-(-v \wedge -w)$ is their least upper bound $v \wedge w$; (iv) for $v \in \mathcal{V}$ there is a smallest element $|v|$ greater than both v and $-v$ — in that case any two elements v, v' have least upper bound $v \vee v' = (v + v' + |v - v'|)/2$ and greatest lower bound $v \wedge v' = (v + v' - |v - v'|)/2$; (v) for any $v \in \mathcal{V}$ there is a least element v_+ exceeding both v and 0, called the ***positive part*** of v — in that case any two elements v, v' have least upper bound $v \vee v' = (v - v')_+ + v'$ and greatest lower bound $v \wedge v' = v' - (v' - v)_+$; (vi) for any $v \in \mathcal{V}$ there is a least element v_- exceeding both $-v$ and 0, called the *negative part* of v — in that case any two elements v, v' have least upper bound $v \vee v' = (v' - v)_- + v'$ and greatest lower bound $v \wedge v' = v' - (v - v')_-$.

Exercise 3.5* Suppose $(\mathcal{V}, \leq)$ is a vector lattice. Prove the following formulae: $(v \wedge v') + w = (v+w) \wedge (v'+w)$; $(v \vee v') + w = (v+w) \vee (v'+w)$; $r(v \wedge v') = (rv) \vee (rv')$ if $r \in \mathbb{R}_+$; $r(v \vee v') = (rv) \vee (rv')$ if $r \in \mathbb{R}_+$; $-(v \wedge v') = (-v) \wedge (-v')$; $-(v \vee v') = (-v) \vee (-v')$; $v + w = v \wedge w + v \vee w$ $v = v_+ - v_-$; $|v| = v_+ + v_-$; $v_+ \wedge v_- = 0$; $v_+ = (|v| + v)/2$; $v_- = (|v| - v)/2$; $v \vee v' = (v + v' + |v - v'|)/2$; $v \wedge v' = (v + v' - |v - v'|)/2$; $|v + v'| \leq |v| + |v'|$; $|rv| = |r| \cdot |v|$.

Exercise 3.6* The compatibility of the order with the linear structure can be expressed in terms of the set of positive elements alone. In order to discuss this, a few elementary notions are required. A subset $\mathcal{C}$ of a real vector space is a ***cone*** if it is closed under multiplication with positive scalars: $\mathbb{R}_+ \cdot \mathcal{C} \subset \mathcal{C}$. A cone $\mathcal{C}$ is ***pointed*** if $\mathcal{C} \cap (-\mathcal{C}) \subseteq \{0\}$. A subset $\mathcal{C}$ of a real vector space is ***convex*** if it contains, with any two elements f, f', the whole line segment from f to f':

$$t \cdot f + (1 - t) \cdot f' \in \mathcal{C} \qquad \text{for } 0 \leq t \leq 1 .$$

(i) Let $(\mathcal{V}, \leq)$ be an ordered vector space over the reals. The set $\mathcal{V}_+ \stackrel{\text{def}}{=} \{v \in \mathcal{V} : v \geq 0\}$ then is a pointed convex cone. It is called the *order cone* of $(\mathcal{V}, \leq)$.
(ii) Conversely, suppose a pointed convex cone $\mathcal{C} \subset \mathcal{V}$ is given, and use it to define an order $\leq$ by $v \leq v' \iff v' - v \in \mathcal{C}$. Show that this order is compatible with the vector space structure and that its order cone is $\mathcal{C}$.

Lemma 3.7 *The vector lattice $(\mathcal{V}, \leq)$ is order complete if every non–void subset of positive elements that is closed under taking finite suprema has a least upper bound.*

Proof. Suppose $\mathcal{S} \subset \mathcal{V}$ is non–void and order bounded above. We want to show that $\mathcal{S}$ has a least upper bound. Let $v_0 \in \mathcal{S}$ and consider the set $\mathcal{S} - v_0 = \{v - v_0 : v \in \mathcal{S}\}$. Since the translation $v \mapsto v - v_0$ is an order isomorphism, $w \in \mathcal{V}$ is a (least) order bound for $\mathcal{S}$ if and only if $w - v_0$ is a (least) order bound for $\mathcal{S} - v_0$. Now $\mathcal{S} - v_0$ contains the zero vector and thus has the same upper bounds as the set $(\mathcal{S} - v_0)_+$ of its positive elements. If among the latter there is a least element, then $\mathcal{S}$ will have a least upper bound. In other words, we may assume that $\mathcal{S}$ contains only positive elements. Consider the family of vectors in $\mathcal{V}$

that are suprema of finite subfamilies of $\mathcal{S}$. This family is evidently closed under taking finite suprema and has the same upper bounds as $\mathcal{S}$. By assumption, there is a smallest one among them. We conclude that the original subset $\mathcal{S}$ has a least upper bound $\bigvee \mathcal{S}$.
Now if $\mathcal{S}$ is order bounded from below, then $-\mathcal{S}$ is order bounded from above and has a least upper bound $\bigvee -\mathcal{S}$. It is then easy to see that $-\bigvee -\mathcal{S}$ is a greatest lower bound of $\mathcal{S}$. ∎

Theorem 3.8 *If the mean* $\| \ \|^*$ *is strictly increasing*[2] *then* L^p *is an order complete Banach lattice, for* $1 \leq p < \infty$. *So is* L^∞, *provided the mean is* σ*–finite.*

Proof. We know from Proposition 3.1 that L^p is a Banach space, and leave it as an exercise to show that its order makes it a Banach lattice. We prove here only the hard part, that it is order complete. By Lemma 3.7 we need to show only that a non–void set $\mathcal{G} \subset L^p_+$ that is closed under taking finite suprema and is order bounded above has a least upper bound.
We start with the case $1 \leq p < \infty$. Let

$$r = \sup\{\|\dot\gamma\|_p : \dot\gamma \in \mathcal{G}\} .$$

This number is finite, since it is majorized by the p–mean of any upper bound for $\mathcal{G}$. There is a sequence $(\dot\gamma_n)$ in $\mathcal{G}$ with $\|\dot\gamma_n\|_p \uparrow r$. We may clearly choose $\dot\gamma_n$ and $\gamma_n \in \dot\gamma_n$ so that $\gamma_{n+1} \geq \gamma_n \quad \forall\, n$. By the Monotone Convergence Theorem the pointwise limit $g = \lim \gamma_n$ is a limit in p–mean. We claim that the class $\dot g$ is the sought–after least upper bound of $\mathcal{G}$.
First, $\dot g$ is an upper bound: since $\|\dot\gamma \vee \dot\gamma_n\|_p \leq r$ for $\dot\gamma \in \mathcal{G}$, we have the same inequality for the limit; i.e., $\|\dot\gamma \vee \dot g\|_p \leq r = \|g\|_p$. Since $\| \ \|_p$ is strictly increasing (Theorem 1.6 (ii)), $\dot\gamma \leq \dot g$.
Second, if $\dot g'$ is any upper order bound, then so is $\dot g'' = \dot g \wedge \dot g'$. Since $\dot g'' \leq \dot g$, we have $\|\dot g''\|_p \leq r$. since $\dot g''$ is an upper bound, we have $\|\dot g''\|_p \geq r$. Thus $\|\dot g''\|_p = r$. The strict monotonicity of the mean $\| \ \|_p$ implies that $\dot g'' = \dot g$, which means $\dot g' \geq \dot g$: $\dot g$ is, indeed, the least upper bound.
This argument does not apply in the case $p = \infty$, inasmuch as $\| \ \|_\infty$ is not strictly increasing. We can, however, reduce this case to the previous one. Let $\{A_\alpha\}$ be a collection of mutually disjoint non–negligible integrable sets that cover the whole space. Since the mean is assumed to be σ–finite, there clearly exists one. The classes $\mathcal{G}_\alpha = \{\dot\gamma \cdot \dot A_\alpha : \ \dot\gamma \in \mathcal{G}\}$ belong to L^1 and are order bounded in L^1. $\mathcal{G}_\alpha$ has a least order bound $\dot g_\alpha$ in L^1. Let $g_\alpha \in \dot g_\alpha$ be a representative that vanishes off A_α, and set $g = \sum_\alpha g_\alpha$. This sum is measurable by the Localization Principle III.3.5, and its class $\dot g$ in L^∞ is the least order bound of $\mathcal{G}$ in L^∞. The details are left to the reader. ∎

Exercise 3.9 None of the spaces $\mathcal{L}^p[\mathbb{R}]$ of *functions* is order complete.

[2] This assumption is actually superfluous, see Exercise 4.10 on p. 118.

The proof of Theorem 3.8 yields some more information:

Corollary 3.10 *Let $1 \le p \le \infty$, and let $\mathcal{G} \subset \mathrm{L}^p$ be an increasingly directed and p–mean bounded set. Then $\mathcal{G}$ is order bounded and has a least upper bound $\dot{g}$. In the cases $1 \le p < \infty$, $\dot{g}$ is the p–mean limit of an increasing sequence in $\mathcal{G}$.*

Exercise 3.11 Let $1 \le p < \infty$, and let $\mathcal{G} \subset \mathrm{L}^p$ be increasingly directed and order bounded above. An upper bound g of $\mathcal{G}$ is the least upper bound iff it belongs to the p–mean closure of $\mathcal{G}$. Give an example showing that this is false in the case $p = \infty$.

Exercise 3.12* Let $p, q, r \ge 1$ be such that $\frac{1}{p} + \frac{1}{q} = \frac{1}{r}$. Let $\mathcal{G} \subset \mathrm{L}^p$ be order bounded with least upper bound g, let $f \in \mathrm{L}^q_+$, and set $f \cdot \mathcal{G} = \{f \cdot g : g \in \mathcal{G}\}$.

(i) $f \cdot \mathcal{G}$ is order bounded in L^r and has least upper bound $f \cdot g$ there.

(ii) Suppose $\mathcal{G}$ is increasingly directed and that $r < \infty$. There exists an increasing sequence (g_n) in $\mathcal{G}$ such that $f \cdot g_n \to f \cdot g$ in r–mean.

Exercise 3.13 A subset of L^p that is order bounded from below has a greatest lower bound.

Exercise 3.14 Let $\| \ \|^*$ be the Daniell mean for counting measure on some set S (see Example II.2.22). The spaces $\mathcal{L}^p$ and L^p coincide. If $S = N$ then $\mathcal{L}^p$ is the space ℓ^p of Exercise I.7.11. If S is uncountable then $\| \ \|^*$ is not σ–finite; nevertheless, L^∞ is order complete.

IV.4 Linear Functionals

Seminormed vector spaces occur with great frequency. As was pointed out before, the analyst will often attack a problem by devising a seminormed space within which it can be well formulated. In fact, this was the way we solved the extension problem for elementary integrals, namely, by the introduction of the Daniell seminorm $\| \ \|^*$. The last sections exhibited a whole slew $\{\mathcal{L}^p : 1 \le p \le \infty\}$ of them. One often has occasion to consider several seminormed spaces $\mathcal{L}, \mathcal{L}', \ldots$ simultaneously. In order to avoid confusion, it is now customary to denote the seminorm on $\mathcal{L}$ by $\| \ \|_{\mathcal{L}}$, that on $\mathcal{L}'$ by $\| \ \|_{\mathcal{L}'}$, etc. In this parlance, $\| \ \|_p = \| \ \|_{\mathcal{L}^p}$ on $\mathcal{L}^p$ and $\| \ \|_p = \| \ \|_{\mathrm{L}^p}$ on L^p.

A major tool, in turn, with which to study the seminormed spaces so arising are their bounded linear functionals. A ***linear functional*** on a real vector space $\mathcal{L}$ is a linear map $f^* : \mathcal{L} \to \mathbb{R}$. It is customary to denote the value of f^* at the point $f \in \mathcal{L}$ by $\langle f^* | f \rangle$. To say that f^* is linear means that for all $f, g \in \mathcal{L}$ and all $r, s \in \mathbb{R}$

$$\langle f^* | r \cdot f + s \cdot g \rangle = r \cdot \langle f^* | f \rangle + s \cdot \langle f^* | g \rangle \ .$$

To say that f^* is ***bounded*** means that there is an $r \in \mathbb{R}$ with

$$|\langle f^* | f \rangle| \le r \cdot \|f\|_{\mathcal{L}} \qquad \forall\, f \in \mathcal{L} \ .$$

Let $\|f^*\|_{\mathcal{L}^*}$ denote the infimum of the numbers r that satisfy this inequality. It is clearly one of them:

$$\big|\langle f^*|f\rangle\big| \le \|f^*\|_{\mathcal{L}^*} \cdot \|f\|_{\mathcal{L}} \qquad \forall\, f \in \mathcal{L}\,,$$

and $\|f^*\|_{\mathcal{L}^*}$ is the smallest number for which this inequality holds. The collection of all bounded linear functionals on $\mathcal{L}$ is called the ***dual*** of $\mathcal{L}$ and is denoted by $\mathcal{L}^*$, and $\|f^*\|_{\mathcal{L}^*}$ is the ***norm of the linear functional*** f^*. This name has to be justified, of course, by establishing the requisite properties of a norm for $\|\ \|_{\mathcal{L}^*}$.

The main result of this section is that a seminormed vector space admits plenty of bounded linear functionals. We start with some observations about their collection $\mathcal{L}^*$ and the functional $\|\ \|_{\mathcal{L}^*}$ on it. First, it is easy to see that $\mathcal{L}^*$ forms a vector space under the usual pointwise operations: $r \cdot f^* + s \cdot g^*$ is the linear functional whose value at $f \in \mathcal{L}$ is $r \cdot \langle f^*|f\rangle + s \cdot \langle g^*|f\rangle$. If f^* and g^* are bounded then so is $r \cdot f^* + s \cdot g^*$, because

$$\big|\langle r \cdot f^* + s \cdot g^*|f\rangle\big| \le |r| \cdot \big|\langle f^*|f\rangle\big| + |s| \cdot \big|\langle g^*|f\rangle\big| \le \Big(|r| \cdot \|f^*\|_{\mathcal{L}^*} + |s| \cdot \|g^*\|_{\mathcal{L}^*}\Big) \cdot \|f\|_{\mathcal{L}}\,.$$

This inequality shows also that $\|\ \|_{\mathcal{L}^*}$ is subadditive. Its absolute–homogeneity is left for the reader to establish. Thus $\|\ \|_{\mathcal{L}^*}$ is a seminorm. It is, in fact, a norm, for if $\|f^*\|_{\mathcal{L}^*} = 0$ then f^* is clearly the zero linear functional.

Proposition 4.1 *$(\mathcal{L}^*, \|\ \|_{\mathcal{L}^*})$ is a Banach space.*

Proof. Only the completeness needs to be established. Now if (f_n^*) is a Cauchy sequence in $\mathcal{L}^*$, that is to say, if

$$\|f_n^* - f_m^*\|_{\mathcal{L}^*} \xrightarrow[m,n\to\infty]{} 0\,,$$

then the sequence $(\langle f_n^*|f\rangle)$ of reals is Cauchy for every $f \in \mathcal{L}$. There exists therefore

$$\langle f^*|f\rangle = \lim_{n\to\infty} \langle f_n^*|f\rangle\,.$$

It is reasonable to write the limit in the form $\langle f^*|f\rangle$, since it is evidently linear in $f \in \mathcal{L}$. Now by Exercise I.7.2 on p. 28,

$$\Big|\|f_n^*\|_{\mathcal{L}^*} - \|f_m^*\|_{\mathcal{L}^*}\Big| \le \|f_n^* - f_m^*\|_{\mathcal{L}^*} \xrightarrow[m,n\to\infty]{} 0\,,$$

so the sequence $(\|f_n^*\|_{\mathcal{L}^*})$ has a limit. From

$$\big|\langle f^*|f\rangle\big| = \lim_n \big|\langle f_n^*|f\rangle\big| \le \lim_n \|f_n^*\|_{\mathcal{L}^*} \cdot \|f\|_{\mathcal{L}}$$

we deduce that f^* is bounded. Next,

$$\big|\langle f^* - f_n^*|f\rangle\big| = \lim_{N\to\infty} \big|\langle f_N^* - f_n^*|f\rangle\big| \le \Big(\sup_{N\ge n} \|f_N^* - f_n^*\|_{\mathcal{L}^*}\Big) \cdot \|f\|_{\mathcal{L}}\,.$$

Given an $\epsilon > 0$ we can make the first factor on the right as small as ϵ by the choice of n, and conclude that $\|f^* - f_n^*\|_{\mathcal{L}^*} \xrightarrow[n\to\infty]{} 0$: the sequence (f_n^*) converges to f^* in the norm of $\mathcal{L}^*$. ∎

Exercise 4.2* Let f^* be a linear functional on the seminormed space $(\mathcal{L}, \| \ \|)$. f^* is bounded if and only if it is continuous at any point, and then it is continuous at every point.

The Hahn–Banach Theorem

For all we know so far $\mathcal{L}^*$ consists only of the zero linear functional. We now show that it is rich. It is convenient to consider at the outset a real vector space $\mathcal{L}$ equipped with a ***gauge***. This is a map $\mathfrak{p} : \mathcal{L} \to \mathbb{R}_+$

that is subadditive: $\mathfrak{p}(f+g) \le \mathfrak{p}(f) + \mathfrak{p}(g)$ $\quad \forall\, f, g \in \mathcal{L}$

and positive–homogeneous: $\mathfrak{p}(r \cdot f) = r \cdot \mathfrak{p}(f)$ $\quad \forall\, 0 \le r \in \mathbb{R}_+$.

In other words, $\mathfrak{p}$ misses being a seminorm by being merely positive–homogeneous rather than absolute–homogeneous—this generality is used on page 117.

Theorem 4.3 (Hahn–Banach) *Let $\mathfrak{p}$ be a gauge on the real vector space $\mathcal{L}$ and f^* a linear functional whose domain is a linear subspace D_{f^*} of $\mathcal{L}$ and that satisfies*

$$\langle f^* | f \rangle \le \mathfrak{p}(f) \qquad \forall\, f \in D_{f^*} .$$

*There exists an **extension** to $\mathcal{L}$ that is still majorized by $\mathfrak{p}$; that is to say, there exists a linear functional F^* that is defined on all of $\mathcal{L}$, coincides with f^* on D_{f^*}, and satisfies*

$$\langle F^* | f \rangle \le \mathfrak{p}(f) \qquad \forall\, f \in \mathcal{L} .$$

Proof. We shall simply extend f^* to larger and larger subspaces of $\mathcal{L}$. If $D_{f^*} = \mathcal{L}$ there is nothing to prove. In the opposite case there is a $g_0 \in \mathcal{L}$ that does not belong to D_{f^*}. Set

$$D_{g^*} = \{ f + r \cdot g_0 : f \in D_{f^*}, r \in \mathbb{R} \} .$$

This is a linear subspace of $\mathcal{L}$ properly containig D_{f^*}. We want to define a linear map $g^* : D_{g^*} \to \mathbb{R}$ that coincides with f^* on D_{f^*} and satisfies

$$\langle g^* | f \rangle \le \mathfrak{p}(g) \quad \forall\, g \in D_{g^*} . \tag{4.1}$$

The linearity forces g^* to be of the form

$$\langle g^* | f + r \cdot g_0 \rangle = \langle f^* | f \rangle + r \cdot \gamma \quad \text{for } f \in D_{f^*} \text{ and } r \in \mathbb{R} , \tag{4.2}$$

where $\gamma = \langle g^*|g_0\rangle$. For any choice of γ, Equation (4.2) clearly defines a linear extension of f^* to D_{g^*}. The question remains whether γ can be chosen so that Inequality (4.1) is satisfied. To see that it can observe that for any $f_1, f_2 \in D_{f^*}$

$$\begin{aligned}\langle f^*|f_1\rangle + \langle f^*|f_2\rangle &= \langle f^*|f_1 + f_2\rangle \\ &\le \mathfrak{p}(f_1 + f_2) = \mathfrak{p}\big((f_1 - g_0) + (f_2 + g_0)\big) \\ &\le \mathfrak{p}(f_1 - g_0) + \mathfrak{p}(f_2 + g_0) ,\end{aligned}$$

and therefore

$$\langle f^*|f_1\rangle - \mathfrak{p}(f_1 - g_0) \le \mathfrak{p}(f_2 + g_0) - \langle f^*|f_2\rangle .$$

There exists therefore a number γ so that

$$\sup_{f_1\in D_{f^*}} \langle f^*|f_1\rangle - \mathfrak{p}(f_1 - g_0) \le \gamma \le \inf_{f_2\in D_{f^*}} \mathfrak{p}(f_2 + g_0) - \langle f^*|f_2\rangle .$$

We fix such a number γ and define g^* on all of D_{g^*} by Equation (4.2). With this choice of $\gamma = \langle g^*|g_0\rangle$ Inequality (4.1) is satisfied. Indeed, let $g = f + rg_0$ be an element of D_{g^*}. If $r = 0$ then Inequality (4.1) is true by assumption on f^*. If $r > 0$ then for $g = f + r \cdot g_0 \in D_{g^*}$

$$\begin{aligned}\langle g^*|g\rangle &= \langle f^*|f\rangle + r\cdot\gamma = r\cdot\Big(\langle f^*|f/r\rangle + \gamma\Big) \\ &\le r\cdot\Big(\langle f^*|f/r\rangle + \big\{\mathfrak{p}(f/r + g_0) - \langle f^*|f/r\rangle\big\}\Big) \\ &= r\cdot\mathfrak{p}(f/r + g_0) = \mathfrak{p}(g) .\end{aligned}$$

If on the other hand $r < 0$ then

$$\begin{aligned}\langle g^*|g\rangle &= \langle f^*|f\rangle + r\cdot\gamma = \langle f^*|f\rangle - |r|\cdot\gamma = |r|\cdot\Big(\langle f^*|f/|r|\rangle - \gamma\Big) \\ &\le |r|\cdot\Big(\langle f^*|f/|r|\rangle - \big\{\langle f^*|f/|r|\rangle - \mathfrak{p}(f/|r| - g_0)\big\}\Big) \\ &= |r|\cdot\mathfrak{p}(f/|r| - g_0) = \mathfrak{p}(f - |r|g_0) = \mathfrak{p}(f + rg_0) = \mathfrak{p}(g) .\end{aligned}$$

We have succeeded "to extend f^* to a subspace of one more dimension." The idea is to keep going until all of $\mathcal{L}$ is exhausted. This has to be made precise. To this end we consider all pairs (D_{g^*}, g^*), where D_{g^*} is a subspace of $\mathcal{L}$ containing D_{f^*} and g^* a linear functional on D_{g^*} that satisfies Inequality (4.1) and coincides with f^* on D_{f^*}. We define a partial order $\preceq$ on such pairs as follows:

$$(D_{g_1^*}, g_1^*) \preceq (D_{g_2^*}, g_2^*)$$

if $D_{g_1^*} \subset D_{g_2^*}$ and g_2^* agrees with g_1^* on $D_{g_1^*}$. According to the Hausdorff Maximal Principle there is a maximal linearly ordered chain $(D_{g_\alpha^*}, g_\alpha^*)$. Set $D_{F^*} = \bigcup_\alpha D_{g_\alpha^*}$. This is a linear subspace of $\mathcal{L}$. If $f \in D_{F^*}$, say $f \in D_{g_\alpha^*}$, we set $\langle F^*|f\rangle = \langle g_\alpha^*|f\rangle$.

This is evidently independent of the α chosen. F^* is the desired extension of f^*. For if it were not, that is to say if D_{F^*} were not the whole space $\mathcal{L}$, then we would apply the construction of the first part of the proof and arrive at an extension of F^* properly $\preceq$–exceeding (D_{F^*}, F^*); this would contradict the maximality of the chain.

Corollary 4.4 *Let* $(\mathcal{L}, \| \ \|_{\mathcal{L}})$ *be a seminormed space and* $f_0 \in \mathcal{L}$. *There exists a linear functional* f^* *of norm* $\|f^*\|_{\mathcal{L}^*} = 1$ *with* $\langle f^* | f_0 \rangle = \|f_0\|_{\mathcal{L}}$.

Proof. Choose $\mathfrak{p} = \| \ \|_{\mathcal{L}}$. On the one–dimensional subspace $D_{f_0^*} = \mathbb{R} \cdot f_0$ define f_0^* by $\langle f_0^* | r f_0 \rangle = r \|f_0\|_{\mathcal{L}}$. f_0^* evidently satisfies $\langle f_0^* | f \rangle \le \|f\|_{\mathcal{L}}$ on $D_{f_0^*}$. There is an extension f^* to all of $\mathcal{L}$ with $\langle f^* | f_0 \rangle = \langle f_0^* | f_0 \rangle = \|f_0\|_{\mathcal{L}}$ that satisfies the same inequality $\langle f^* | f \rangle \le \|f\|_{\mathcal{L}}$, for all $f \in \mathcal{L}$. If $f \in \mathcal{L}$ is such that $\langle f^* | f \rangle \ge 0$ then clearly $\big|\langle f^* | f \rangle\big| \le \|f\|_{\mathcal{L}}$ holds. If this number is negative, then

$$\big|\langle f^* | f \rangle\big| = \langle f^* | -f \rangle \le \| -f \|_{\mathcal{L}} = \|f\|_{\mathcal{L}} .$$

This inequality says $\|f^*\|_{\mathcal{L}^*} \le 1$.

Corollary 4.5 *Let* $(\mathcal{L}, \| \ \|_{\mathcal{L}})$ *be a seminormed space. For* $f \in \mathcal{L}$ *and* $f^* \in \mathcal{L}^*$

$$\|f\|_{\mathcal{L}} = \max\Big\{\big|\langle f^* | f \rangle\big| : \ f^* \in \mathcal{L}^*, \ \|f^*\|_{\mathcal{L}^*} \le 1\Big\} ,$$

and

$$\|f^*\|_{\mathcal{L}^*} = \sup\Big\{\big|\langle f^* | f \rangle\big| : \ f \in \mathcal{L}, \ \|f\|_{\mathcal{L}} \le 1\Big\} .$$

Exercise 4.6* Let $(\mathcal{L}, \| \ \|_{\mathcal{L}})$ be a seminormed space. Every element f of $\mathcal{L}$ defines a continuous linear functional $j(f)$ on $\mathcal{L}^*$ *via* $f^* \mapsto \langle f^* | f \rangle$, for $f^* \in \mathcal{L}^*$. In other words, $\langle j(f) | f^* \rangle = \langle f^* | f \rangle$. We obtain in this way a map j from $\mathcal{L}$ into its ***double dual*** $\mathcal{L}^{**}$. Show that j is linear and isometric from $\mathcal{L}$ *into* $\mathcal{L}^{**}$:

$$\|j(f)\|_{\mathcal{L}^{**}} \le \|f\|_{\mathcal{L}} .$$

Show that j is an injection if and only if $\| \ \|_{\mathcal{L}}$ is a norm. In this case we may visualize $\mathcal{L}$ as a subspace of its own double dual.
If j is a surjection then $\mathcal{L}$ is called ***reflexive***. Show that if $\mathcal{L}$ is reflexive then so is $\mathcal{L}^*$.

Positive Linear Functionals

Many of the seminormed spaces we have met are ordered in a natural way. The question arises whether there are ***positive bounded linear functionals***, that is to say, functionals $f^* \in \mathcal{L}^*$ satisfying

$$\mathcal{L} \ni f \ge 0 \ \text{ implies } \ \langle f^* | f \rangle \ge 0 .$$

There are, and this can be established as another application of the Hahn–Banach Theorem. Recall the notion of a convex set from Exercise 3.6 on p. 110: A subset $\mathcal{C}$ of a real vector space $\mathcal{L}$ is ***convex*** if it contains, with any two elements f, f', the whole line segment from f to f':

$$t \cdot f + (1-t) \cdot f' \in \mathcal{C} \qquad \text{for } 0 \leq t \leq 1 .$$

Corollary 4.7 *Suppose* $(\mathcal{L}, \leq, \| \ \|_{\mathcal{L}})$ *is an ordered seminormed vector space. Let* f_0 *be a positive element of* $\mathcal{L}$. *There exists a positive* $f^* \in \mathcal{L}^*$ *with norm* $\|f^*\|_{\mathcal{L}^*} \leq 1$ *and* $\langle f^* | f_0 \rangle = \|f_0\|_{\mathcal{L}}$.

Proof. Let $\mathcal{P}$ be the algebraic sum of the unit ball and the cone of negative elements:

$$\mathcal{P} = \{ f_1 + f_2 \ : \ \|f_1\|_{\mathcal{L}} \leq 1, \ f_2 \leq 0 \} .$$

This is clearly a convex set. Observe also that $f \in \mathcal{P}$ implies $t \cdot f \in \mathcal{P}$ as long as $0 \leq t \leq 1$. Let $\mathfrak{p}$ be the ***Minkowski functional*** of $\mathcal{P}$:

$$\mathfrak{p}(f) = \inf \{ s \in \mathbb{R}_+ : \ f \in s \cdot \mathcal{P} \} \qquad f \in \mathcal{L} .$$

This is finite, since $f \in \|f\| \cdot \mathcal{P}$ and thus

$$\mathfrak{p}(f) \leq \|f\|_{\mathcal{L}} .$$

$\mathfrak{p}$ is positive–homogeneous, since $\{s \geq 0 : rf \in s\mathcal{P}\} = r\{s \geq 0 : f \in s\mathcal{P}\}$ for $r \geq 0$. It is also subadditive. Indeed, there arbitrarily small $\epsilon_1, \epsilon_2 > 0$ with $f_1 \in (\mathfrak{p}(f_1) + \epsilon_1)\mathcal{P}$ and $f_2 \in (\mathfrak{p}(f_2) + \epsilon_2)\mathcal{P}$. That is to say, there are $f_1', f_2' \in \mathcal{P}$ with $f_i = (\mathfrak{p}(f_i) + \epsilon_i) \cdot f_i'$. Now

$$\begin{aligned} f_1 + f_2 = &(\mathfrak{p}(f_1) + \mathfrak{p}(f_2) + \epsilon_1 + \epsilon_2) \times \\ &\times \left[\frac{\mathfrak{p}(f_1) + \epsilon_1}{\mathfrak{p}(f_1) + \mathfrak{p}(f_2) + \epsilon_1 + \epsilon_2} \cdot f_1' + \frac{\mathfrak{p}(f_2) + \epsilon_2}{\mathfrak{p}(f_1) + \mathfrak{p}(f_2) + \epsilon_1 + \epsilon_2} \cdot f_2' \right] . \end{aligned}$$

The second factor belongs to $\mathcal{P}$ because of convexity, and therefore

$$\mathfrak{p}(f_1 + f_2) \leq \mathfrak{p}(f_1) + \mathfrak{p}(f_2) + \epsilon$$

for arbitrarily small $\epsilon > 0$. This establishes the subadditivity. Note that $\mathfrak{p}$ is not absolute–homogeneous — it was to cover this situation that gauges were introduced on page 114. However, $\mathfrak{p}$ is not too far from the norm:

$$\mathfrak{p}(f) = \|f\|_{\mathcal{L}} \ \text{ for } \ f \geq 0 .$$

The inequality $\mathfrak{p}(f) \le \|f\|_{\mathcal{L}}$ is known from above. For the converse inequality let $f \in \mathcal{L}_+$ have $\mathfrak{p}(f) < s$. There are then f_1 with $\|f_1\|_{\mathcal{L}} \le 1$ and $f_2 \in \mathcal{L}_+$ such that $f = s(f_1 - f_2)$. We have $0 \le f \le sf_1$ and, due to the solidity of the seminorm, $\|f\|_{\mathcal{L}} \le \|sf_1\|_{\mathcal{L}} \le s$.
We proceed now as in Corollary 4.4. On the one–dimensional subspace $D_{f_0^*} = \mathbb{R}\cdot f_0$ we define f_0^* by $\langle f_0^* | r f_0 \rangle = r\|f_0\|_{\mathcal{L}}$. f_0^* satisfies $\langle f_0^* | f \rangle \le \mathfrak{p}(f)$ on $D_{f_0^*}$. For if $D_{f_0^*} \ni f = rf_0$ with $r \ge 0$ then

$$\langle f_0^* | f \rangle = r\|f_0\|_{\mathcal{L}} = r\mathfrak{p}(f_0) = \mathfrak{p}(f) \;;$$

and if $r < 0$ then $\langle f_0^* | f \rangle \le 0 \le \mathfrak{p}(f)$. There is an extension f^* to all of $\mathcal{L}$ with $\langle f^* | f_0 \rangle = \langle f_0^* | f_0 \rangle = \|f_0\|_{\mathcal{L}}$ that satisfies the same inequality $\langle f^* | f \rangle \le \mathfrak{p}(f) \le \|f\|_{\mathcal{L}}$ for all $f \in \mathcal{L}$. We conlude as in Corollary 4.4 that $\|f^*\|_{\mathcal{L}^*} \le 1$. Now if $f \le 0$ then $\mathfrak{p}(f) = 0$ and thus $\langle f^* | f \rangle \le 0$: f^* is positive, as desired. ∎

Exercise 4.8 Let K be a compact Hausdorff space. For every positive continuous function ϕ on K there exists a positive Radon measure μ with $\mu(\phi) = \|\phi\|_u$.

Exercise 4.9 The dual of a seminormed vector lattice is an order complete Banach lattice.

Exercise 4.10* Let $\| \;\|^*$ be a mean on $\mathcal{E}$ and $\epsilon > 0$. There exists a strictly increasing mean $\| \;\|^\sharp$ so that for all $f : S \to \overline{\mathbb{R}}$

$$\|f\|^* \le \|f\|^\sharp \le (1+\epsilon)\|f\|^* \tag{4.3}$$

IV.5 The Dual of L^p

From now on, $\| \;\|^*$ is the Daniell mean for some σ–additive elementary integral $\int$. We set ourselves the task to identify the bounded linear functionals on $\mathcal{L}^p$, for $1 \le p < \infty$.

Example 5.1 Let $1 \le p < \infty$ and $\mathcal{L} = \mathcal{L}^p$, let p' be the conjugate exponent —defined by $1/p + 1/p' = 1$—and let $\dot{g} \in \mathrm{L}^{p'}$. Consider the number

$$\langle \dot{g} | f \rangle \stackrel{\text{def}}{=} \int g\cdot f\,, \qquad f \in \mathcal{L}^p\,,\; g \in \dot{g}\,. \tag{5.1}$$

By Hölder's inequality, this number is finite. It clearly does not depend on the representative $g \in \dot{g}$ and is linear in f. The map $f \mapsto \langle \dot{g} | f \rangle$ is therefore a linear functional on $\mathcal{L}^p$. Let us call it $\dot{g}^*$. $\dot{g}^*$ is bounded; in fact, Exercise 1.3 gives

$$\|\dot{g}^*\|_{(\mathcal{L}^p)^*} = \sup\{\langle \dot{g} | f \rangle : \|f\|_p \le 1\} = \|\dot{g}\|_{p'}\,. \tag{5.2}$$

The upshot of all this is that every class $\dot{g}$ in $\mathrm{L}^{p'}$ acts as a continuous linear functional on $\mathcal{L}^p$ *via* the *pairing* (5.1), and the norm of that linear functional in the dual $(\mathcal{L}^p)^*$ is equal to the norm of $\dot{g}$ in $\mathrm{L}^{p'}$.

Exercise 5.2* The map $\dot{g} \mapsto \dot{g}^*$ from $\mathrm{L}^{p'}$ to $(\mathcal{L}^p)^*$ is linear, injective, order preserving, and isometric.

Exercise 5.3* Let $\mathbf{g}^*$ be a bounded linear functional on $\mathcal{L}^p$, not necessarily of the form (5.1). Then $\mathbf{g}^*$ is positive iff $\langle \mathbf{g}^* | A \rangle \ge 0$ for all sets A. Next let A be a measurable set. Show that $f \mapsto \langle \mathbf{g}^* | f \cdot A \rangle$ is again a bounded linear functional on $\mathcal{L}^p$, positive if $\mathbf{g}^*$ is.

We show now that the map $\dot{g} \mapsto \dot{g}^*$ from $\mathrm{L}^{p'}$ to $(\mathcal{L}^p)^*$ is surjective, in other words, that all bounded linear functionals on $\mathcal{L}^p$ are of the form (5.1).

Proposition 5.4 *Let $1 < p < \infty$. For every bounded linear functional $\mathbf{g}^*$ on $\mathcal{L}^p$ there exists a class $\dot{g} \in \mathrm{L}^{p'}$, $(\frac{1}{p} + \frac{1}{p'} = 1)$, such that for all $f \in \mathcal{L}^p$ and $g \in \dot{g}$*

$$\langle \mathbf{g}^* | f \rangle = \int g \cdot f \qquad \Big(= \langle \dot{g}^* | f \rangle \Big) . \tag{5.3}$$

The same is true for $p = 1$, provided $\| \ \|^*$ is σ-finite.

Proof. The bounded linear functional $\mathbf{g}^*$ is rather an abstract thing, and the reader may well wonder how we will conjure up a "real" function g or class $\dot{g}$ to implement it *via* (5.3). To get a clue, consider a function $g \in \mathcal{L}^{p'}$ and let us try to identify the class $\dot{g}_+$ of the positive part of g from the linear functional $\mathbf{g}^* = \dot{g}^*$. To this end note that, for any positive function $f \in \mathcal{L}^p$, $\int g \cdot f \le \int g \cdot [g > 0] \cdot f = \langle \dot{g}^* | f \cdot [g > 0] \rangle$. More generally, for any function $0 \le \gamma \le g_+$,

$$\begin{aligned} \int \gamma \cdot f &= \int \gamma [\gamma > 0] \cdot f \\ &\le \int g[\gamma > 0] \cdot f = \int g \cdot f[\gamma > 0] \\ &= \langle \mathbf{g}^* | f[\gamma > 0] \rangle , \end{aligned}$$

and $\dot{g}_+$ is the largest class of such γ. This little reflection leads us to try to identify $\dot{g}_+$ as the least upper bound in $\mathrm{L}^{p'}$ of the set of classes

$$\mathcal{G} = \Big\{ \dot{\gamma} \in \mathrm{L}^{p'}_+ : \int \gamma \cdot f \le \langle \mathbf{g}^* | f \cdot [\gamma > 0] \rangle \quad \forall \, f \in \mathcal{L}^p_+ \Big\} .$$

The set $\mathcal{G}$ can be defined for an arbitrary continuous linear functional $\mathbf{g}^*$ on $\mathcal{L}^p$. Our strategy will be to prove the following *Claim: $\mathcal{G}$ has a least upper bound $\dot{g}_+ = \dot{g}_+[\mathbf{g}^*]$, and the corresponding linear functional $\dot{g}_+^*$ exceeds $\mathbf{g}^*$.*

From this it is easy to see that actually $\mathbf{g}^* = \dot{g}_+^*$ when $\mathbf{g}^*$ is positive. Indeed, we have then for every $0 \le f \in \mathcal{L}^p$ and $g_+ \in \dot{g}_+$

$$\langle \dot{g}_+^* | f \rangle \ge \langle \mathbf{g}^* | f \rangle \ge \langle \mathbf{g}^* | f \cdot [g_+ > 0] \rangle \ge \int g_+ \cdot f = \langle \dot{g}_+^* | f \rangle ,$$

so that equality must obtain throughout. And if $\mathbf{g}^*$ and $\dot{g}_+^*$ agree on $\mathcal{L}_+^p$ then they agree on all of $\mathcal{L}^p$.

We then apply the claim and its consequence above to the positive linear functional $\dot{g}_+^* - \mathbf{g}^*$. This equals $\dot{g}_-^*$ for some positive class $\dot{g}_- \in \mathrm{L}^{p'}$, so that $\mathbf{g}^* = \dot{g}^*$ with $\dot{g} = \dot{g}_+ - \dot{g}_- \in \mathrm{L}^{p'}$.

Thus all will be done if the claim above is established. We start on it by noting that $\mathcal{G}$ is increasingly directed: if $\dot{\gamma}_1, \dot{\gamma}_2 \in \mathcal{G}$ and $f \in \mathcal{L}_+^p$ then

$$\begin{aligned}\int (\gamma_1 \vee \gamma_2) \cdot f &= \int [\gamma_1 \geq \gamma_2]\cdot\gamma_1\cdot f + \int [\gamma_1 < \gamma_2]\cdot\gamma_2\cdot f \\ &\leq \langle \mathbf{g}^* | f\cdot[\gamma_1 \geq \gamma_2] \cap [\gamma_1 > 0]\rangle + \langle \mathbf{g}^* | f\cdot[\gamma_1 < \gamma_2] \cap [\gamma_2 > 0]\rangle \\ &= \langle \mathbf{g}^* | f\cdot[\gamma_1 \vee \gamma_2 > 0]\rangle \, .\end{aligned}$$

Thus $\dot{\gamma}_1 \vee \dot{\gamma}_2 \in \mathcal{G}$.

Next observe that $\mathcal{G}$ is norm–bounded in $\mathrm{L}^{p'}$ by $\|\mathbf{g}^*\|_{(\mathcal{L}^p)^*}$. Indeed, by Exercise 1.3 on p. 103 we have for every $\dot{\gamma} \in \mathcal{G}$

$$\begin{aligned}\|\gamma\|_{p'} &= \sup\Big\{ \int \gamma f \ : \ 0 \leq f \in \mathcal{L}^p \, , \ \|f\|_p \leq 1 \Big\} \\ &\leq \sup\big\{ \langle \mathbf{g}^* | f[\gamma > 0]\rangle \ : \ 0 \leq f \in \mathcal{L}^p \, , \ \|f\|_p \leq 1 \big\} \\ &\leq \sup\big\{ \langle \mathbf{g}^* | f\rangle \ : \ 0 \leq f \in \mathcal{L}^p \, , \ \|f\|_p \leq 1 \big\} \leq \|\mathbf{g}^*\|_{(\mathcal{L}^p)^*} \, .\end{aligned}$$

If $p' = \infty$ this means that $\mathcal{G}$ is order bounded, and Theorem 3.8 provides a least upper bound $\dot{g}_+$ for $\mathcal{G}$ in L^∞. If $1 < p < \infty$ and consequently $p' < \infty$ then the least upper bound $\dot{g}_+$ of $\mathcal{G}$ in $\mathrm{L}^{p'}$ is provided by Corollary 3.10.

From Exercise 3.12 we know that there exists an increasing sequence $(\dot{\gamma}_n)$ in $\mathcal{G}$ such that $\gamma_n f \uparrow g_+ f$ in 1–mean. By the DCT, $[\gamma_n > 0] \cdot f \to [g_+ > 0] \cdot f$ in p–mean, and therefore

$$\int g_+ f = \lim \int \gamma_n \cdot f \leq \lim \, \langle \mathbf{g}^* | f\cdot[\gamma_n > 0]\rangle = \langle \mathbf{g}^* | f\cdot[g_+ > 0]\rangle \, .$$

This proves that $\mathcal{G}$ contains its least upper bound $\dot{g}_+$.

To finish the claim we must show that the continuous linear functional $\dot{g}_+^* - \mathbf{g}^*$ is *positive*:

$$\langle \dot{g}_+^* - \mathbf{g}^* | f\rangle \geq 0 \quad \forall \, f \in \mathcal{L}_+^p \, . \tag{5.4}$$

To prove this write $\Delta^* = \mathbf{g}^* - \dot{g}_+^*$ and assume, by way of contradiction, that (5.4) is false. There is then a positive $f \in \mathcal{L}^p$ with $\langle \Delta^* | f\rangle > 0$. By Exercise 5.3 there exists an integrable set A with $\langle \Delta^* | A\rangle > 0$. A is clearly not negligible. Archimedes' principle provides a number $r > 0$ so that

$$r \cdot \int A < \langle \Delta^* | A\rangle \, .$$

Let $\mathcal{C}$ a maximal disjoint collection of non–negligible integrable subsets C of A with

$$r \cdot \int C \geq \langle \Delta^* | C \rangle .$$

This collection is countable (Exercise II.7.4), and its union $\bigcup \mathcal{C}$ is not A, not even almost everywhere, since it satisfies again

$$r \cdot \int \bigcup \mathcal{C} \geq \langle \Delta^* | \bigcup \mathcal{C} \rangle .$$

The set $B = A \setminus \bigcup \mathcal{C}$ is non–negligible and has the property that $r \cdot \int S < \langle \Delta^* | S \rangle$ for every one of its non–negligible integrable subsets S. We conclude that

$$r \cdot \int SB \leq \langle \Delta^* | SB \rangle$$

for every integrable set S. Taking linear combinations with positive coefficients, we see that

$$\int f \cdot rB \leq \langle \Delta^* | f \cdot B \rangle \tag{5.5}$$

for positive simple integrable functions. Since these are $\| \; \|_p$–dense in $\mathcal{L}^p_+$ and since both sides of (5.5) are $\| \; \|_p$–mean continuous in f, we conclude that (5.5) for arbitrary positive p–integrable functions. In view of the definition of Δ^*, (5.5) reads

$$\int B \cdot (g_+ + rB) \cdot f \leq \langle \mathbf{g}^* | f \cdot B \rangle \quad \forall \, f \in \mathcal{L}^p_+ .$$

To this we add the inequality

$$\int B^c \cdot (g_+ + rB) \cdot f = \int g_+ \cdot (f B^c) \leq \langle \mathbf{g}^* | f \cdot B^c \cdot [g_+ > 0] \rangle ,$$

which uses the fact that $\dot{g}_+ \in \mathcal{G}$, and obtain

$$\int (g_+ + rB) \cdot f \leq \langle \mathbf{g}^* | f \cdot B \rangle + \langle \mathbf{g}^* | f \cdot B^c \cdot [g_+ > 0] \rangle = \langle \mathbf{g}^* | f \cdot [g_+ + rB > 0] \rangle .$$

This means that $\dot{g}_+ + r\dot{B} \in \mathcal{G}$, which is in plain contradiction to the fact that g_+ is an upper bound for $\mathcal{G}$: (5.4) is true, the claim is established, and the is proposition proved. ∎

Theorem 5.5 (Summary) *The map $\dot{g} \mapsto \dot{g}^*$ is an order preserving isometric linear isomorphism of $\mathrm{L}^{p'}$ onto the dual of $\mathcal{L}^p$, for $1 < p < \infty$, and also for $p = 1$ if the integral is σ–finite. In other words, $(\mathcal{L}^p)^*$ can be identified with $\mathrm{L}^{p'}$:*

$$\Big((\mathcal{L}^p)^*, \| \; \|_{(\mathcal{L}^p)^*} \Big) \simeq \Big(\mathrm{L}^{p'}, \| \; \|_{p'} \Big) .$$

Supplements and Additional Exercises

Complex L^p Spaces. We define, naturally, the space $\mathrm{L}^p_{\mathbb{C}}$ as the closure of the elementary complex–valued functions $\mathcal{E} \otimes \mathbb{C}$ under the seminorm $f \mapsto \|\|f\|\|_p$ when $1 \le p < \infty$, as the space of complex–valued measurable functions with $\|\|f\|\|_\infty < \infty$ for $p = \infty$. We know from Exercise III.3.12 on p. 78 exactly what these spaces are. We might have misgivings about reproducing the results of this section for them, since the order played such an important rôle in the proofs. However, the results not involving order are still true:

Exercise 5.6 Hölder's and Minkowski's inequalities and Exercise 1.3 stay.

Let us associate with every $g \in \mathrm{L}^{p'}_{\mathbb{C}}$ a linear functional g^* on $\mathcal{L}^p_{\mathbb{C}}$ by

$$\langle g^* | f \rangle = \int \overline{g} \cdot f \,.$$

Note the complex conjugation on g; the reason for it will become clear in the supplements to Section 6.

Exercise 5.7 $g \mapsto g^*$ is an antilinear ($(cg)^* = \overline{c} g^*$) isometry of $\mathrm{L}^{p'}_{\mathbb{C}}$ onto $(\mathrm{L}^p_{\mathbb{C}})^*$.

IV.6 The Hilbert space L^2

If $p = 2$ then its conjugate exponent p' equals 2 as well, and Theorem 5.5 identifies L^2 with its own dual. This makes L^2 a space of particular interest. It looks much like the euclidean space with its euclidean norm $\| \ \|_2$, except that it is infinite–dimensional. Its norm comes also from an inner product $\langle \ | \ \rangle$, to wit, $\langle g|f \rangle = \int gf$. Here as there we have $\|f\| = \sqrt{\langle f|f \rangle}$.

Definition 6.1 *A* ***real inner product space*** *is a vector space $\mathcal{H}$ over the reals equipped with a map $\langle \ | \ \rangle : \mathcal{H} \times \mathcal{H} \to \mathbb{R}$, called the* ***inner product****, that has the following properties:*

it is symmetric:	$\langle f\|g \rangle = \langle g\|f \rangle$	$\forall f, g \in \mathcal{H}$,
bilinear:	$\langle rf + sg\|h \rangle = r\langle f\|h \rangle + s\langle g\|h \rangle$	$\forall f, g, h \in \mathcal{H}$, $\forall r, s \in \mathbb{R}$,
and positive:	$\langle f\|f \rangle \ge 0$	$\forall f \in \mathcal{H}$.

Exercise 6.2 Show that $\langle h|rf + sg \rangle = r\langle f|h \rangle + s\langle f|g \rangle$. This together with the second line above explains the name "bilinear."

Since $\langle f|f \rangle$ is always positive, we can take the root and obtain the ***associated seminorm***: $\|f\| = \sqrt{\langle f|f \rangle}$. The name must be justified, of course, by showing that $\| \ \|$ is absolute–homogeneous and subadditive. We shall do that below. In an inner product space we have the ***Cauchy–Schwarz inequality***

$$|\langle g|f\rangle| \leq \|g\| \cdot \|f\| \ . \tag{6.1}$$

To see this consider the polynomial in t

$$p(t) = \|g - tf\|_2 = \|g\|_2 - 2t\langle g|f\rangle + t^2\|f\|_2 \ .$$

It is positive for all t. If $\|f\| = 0$ then this is possible only if $\langle g|f\rangle = 0$ as well, and the desired Inequality (6.1) follows. If $\|f\| > 0$ then we choose $t = \langle g|f\rangle/\|f\|_2$, the point where the parabola $p(t)$ takes its minimal value, and get

$$\|g\|_2 - 2\frac{\langle g|f\rangle^2}{\|f\|_2} + \frac{\langle g|f\rangle^2}{\|f\|_2} \geq 0 \ ,$$

which yields the Cauchy–Schwarz inequality in that case. From this we get ***Minkowski's inequality***

$$\|f + g\| \leq \|f\| + \|g\| \ .$$

Simply take the root in $\|f+g\|_2 = \|f\|_2 + 2\langle f|g\rangle + \|g\|_2 \leq \|f\|_2 + 2\|f\|\cdot\|g\| + \|g\|_2 = (\|f\| + \|g\|)^2$. Since $\| \ \|$ is evidently absolute–homogeneous, it is a seminorm.

Definition 6.3 *A real inner product space $(\mathcal{H}, \langle \ | \ \rangle)$ is a* ***real Hilbert space*** *if the associated seminorm $\| \ \| = \sqrt{\langle \ | \ \rangle}$ is a norm and if $(\mathcal{H}, \| \ \|)$ is complete.*

For instance, the space L^2 of Section 3 is a Hilbert space under the inner product

$$\langle f|g\rangle \stackrel{\text{def}}{=} \int fg \ .$$

Let us show that the elementary geometric properties of a Hilbert space are very similar to those of ordinary euclidean $\mathbb{R}^n$. First a few notions. An element f of the Hilbert space $\mathcal{H}$ is called ***normalized***, or a ***unit vector***, if $\|f\| = 1$. Two elements f, g of $\mathcal{H}$ are called ***perpendicular*** or ***orthogonal*** if $\langle f|g\rangle = 0$. We then write $f \perp g$. A subset $S \subset \mathcal{H}$ is called orthogonal if any two of its members are orthogonal; it is called ***orthonormal*** if it is orthogonal and all its members are normalized. A maximal orthonormal subset of $\mathcal{H}$ is an ***orthonormal basis***.

Exercise 6.4 $\|f + g\|_2^2 = \|f\|_2^2 + \|g\|_2^2$ if and only if $f \perp g$.

Every Hilbert space $\mathcal{H}$ has a basis, by Zorn's lemma. Bases have the same function in Hilbert space as they have in euclidean space:

Theorem 6.5 *Let B be an orthonormal basis of the Hilbert space $\mathcal{H}$. Every element $f \in \mathcal{H}$ can be written uniquely as an infinite sum*

$$f = \sum_{n=1}^{\infty} r_n \cdot e_n , \qquad r_n \in \mathbb{R} , e_n \in B .$$

Furthermore,
$$\|f\|_2^2 = \sum_{n=1}^{\infty} r_n^2 .$$

Proof. Let $B_0 = \{e_1, \ldots, e_N\}$ be a finite subcollection of B of size N and consider the numbers $r_n = \langle e_n | f \rangle$, $n = 1, \ldots, N$, and the element $f_N = \sum_{n=1}^{N} r_n e_n$ of $\mathcal{H}$. Since $f_N \perp f - f_N$ we have $\|f_N\|_2^2 + \|f - f_N\|_2^2 = \|f\|_2^2$, so that

$$\|f_N\|_2^2 = \sum_{n=1}^{N} r_n^2 \leq \|f\|_2^2 .$$

[This is called Bessel's Inequality.] It implies that of the numbers $\langle e | f \rangle$, $e \in B$, at most countably many may be non–zero. Let $\{e_1, e_2, \ldots\} \subset B$ be the basis elements whose inner product with f is non–zero, and set

$$r_n = \langle e_n | f \rangle \quad \text{and} \quad f' = \sum_{n=1}^{\infty} r_n \cdot e_n .$$

This sum converges because of the completeness of $\mathcal{H}$. It easily seen that $f - f'$ is perpendicular to every member of B. If it weren't zero, it could be normalized and adjoined to B, in contradiction to the maximality of this set. Thus $f = f'$. We leave as an exercise the

Corollary 6.6 *With the notation above, we have* ***Parseval's Identity***

$$\|f\|_2^2 = \sum_{n=1}^{\infty} r_n^2 = \sum \{\langle e | f \rangle^2 : e \in B\} .$$

Exercise 6.7 Any two bases of $\mathcal{H}$ have the same cardinality. If $\mathcal{H}$ is separable then every basis is at most countable; in this case $\mathcal{H}$ is linearly isometric with ℓ^2.

A subset of $\mathcal{H}$ is said to ***generate*** $\mathcal{H}$ if its linear span is dense in $\mathcal{H}$.

Exercise 6.8 Suppose $S \subset \mathcal{H}$ generates $\mathcal{H}$ and $\mathcal{H}$ is separable. Use the Gram–Schmid procedure to produce a basis in the linear span of S. Conclude that $\mathrm{L}^2[\lambda]$ has a basis consisting of real–analytic functions.

Exercise 6.9 Let (f_n) be a sequence in the Hilbert space $\mathcal{H}$ whose unordered partial sums form a bounded set:

$$\Big\{\sum_{n \in N} f_n : \mathbb{N} \supset N \text{ finite } \Big\} \text{ is bounded.} \qquad (PSB)$$

Show that $f_n \to 0$. A space in which all sequences with property (PSB) converge to zero is called a ***C–space*** [10]. The spaces L^p all are C–spaces for $1 \leq p < \infty$, $(C_0[\mathbb{R}], \| \ \|_u)$ is not.

Chapter V

Operations on Measures

This chapter deals with products and images of measures, signed measures, distributions of measurable functions and maps, convolution, and interpolation.

V.1 Products of Elementary Integrals

The Lebesgue measure or area of a rectangle in the plane is the product of the Lebesgue measures or lengths of its sides. The integral of a function on the plane can be evaluated by iterated integration. These are simple facts from the Calculus. Analyzing them and generalizing them leads to the notion of a product of two elementary integrals and to Fubini's theorem.

Let X, Y be two sets and $(\mathcal{E}_X, m_X)$, $(\mathcal{E}_Y, m_Y)$ elementary integrals on them. The ***product***

$$(\mathcal{E}, m) = (\mathcal{E}_X \otimes \mathcal{E}_Y, m_X \times m_Y))$$

of these two elementary integrals is an elementary integral on the cartesian product $X \times Y$. The elementary integrands on $X \times Y$ are the finite sums of the form

$$(x,y) \mapsto \phi(x,y) = \sum_{i=1}^{I} \phi_X^{(i)}(x) \cdot \phi_Y^{(i)}(y)\,, \quad \phi_X^{(i)} \in \mathcal{E}_X, \phi_Y^{(i)} \in \mathcal{E}_Y\,. \tag{1.1}$$

Their collection is denoted by $\mathcal{E}_X \otimes \mathcal{E}_Y$ or $\mathcal{E}$ for short. The elementary integral of ϕ as in (1.1) is defined by

$$\begin{aligned} m(\phi) = m_X \times m_Y(\phi) &= \int_{X \times Y} \phi(x,y)\, m(dx,dy) \\ &= \sum_{i=1}^{I} m_X(\phi_X^{(i)}) \cdot m_Y(\phi_Y^{(i)})\,. \end{aligned} \tag{1.2}$$

The collection $\mathcal{E} = \mathcal{E}_X \otimes \mathcal{E}_Y$ of functions of the form (1.1) is plainly a ring of bounded functions on $X \times Y$. It is not, in general, a vector lattice (Exercise I.2.19),

K. Bichteler, *Integration – A Functional Approach*, Modern Birkhäuser Classics,
DOI 10.1007/978-3-0348-0055-6_5, © Birkhäuser Verlag 1998

but let that not bother us for the moment. Let us note, though, that the definitions of both $\mathcal{E} = \mathcal{E}_X \otimes \mathcal{E}_Y$ and of $m = m_X \times m_Y$ are symmetric in X, Y.

A function $\phi \in \mathcal{E}_X \otimes \mathcal{E}_Y$ has different representations of the form (1.1) in general, so it must be shown that the number $m(\phi)$ of (1.2) does not depend on the choice of the representation. To see this, let $\phi \in \mathcal{E}$ be as in (1.1). Then, for fixed $x \in X$, $y \mapsto \phi(x,y)$ is an elementary function of $y \in Y$, and so we may apply m_Y:

$$\int_Y \phi(x,y)\, m_Y(dy) = \sum_{i=1}^{I} \phi_X^{(i)}(x) \cdot m_Y(\phi_Y^{(i)})$$

not only makes sense for every $x \in X$ but produces an elementary function in $\mathcal{E}_X$. We may apply m_X to it, and obtain

$$\int_X \left[\int \phi(x,y)\, m_Y(dy) \right] m_X(dx) = \sum_{i=1}^{I} m_X(\phi_X^{(i)}) \cdot m_Y(\phi_Y^{(i)}) = m(\phi) . \tag{1.3}$$

The left–hand side does not depend on the representation of ϕ, so the definition (1.2) is good. Note that (1.3) says that the product of the elementary integrals can be evaluated by iterated integration. Since the original definition (1.2) is symmetric in X, Y, we can evaluate $m_X \times m_Y(\phi)$ also the other way around, by iterated integration first with m_X and then with m_Y. Our ultimate goal is to prove, under the name *Fubini's theorem*, that the integral of every $m_X \times m_Y$–integrable function on $X \times Y$ can be evaluated by iterated integration either way around.

Example 1.1 (Products of Additive Set Functions) Let $\mathcal{A}_X, \mathcal{A}_Y$ be rings of sets on X, Y, respectively, and let μ_X, μ_Y be measures on them. The ***product*** $\mathcal{A}_X \otimes \mathcal{A}_Y$ is the ring of sets on $X \times Y$ generated by the ***rectangles***

$$A \times B , \qquad\qquad A \in \mathcal{A}_X, B \in \mathcal{A}_Y .$$

The ***product measure*** is defined on such rectangles by

$$\mu_X \times \mu_Y(A \times B) = \mu_X(A) \cdot \mu_Y(B) . \tag{$*$}$$

There is a unique additive extension of $\mu_X \times \mu_Y$ to all of $\mathcal{A}_X \otimes \mathcal{A}_Y$; it is again denoted by $\mu_X \times \mu_Y$. This is seen easiest by considering the product $m_X \times m_Y$ of the linear extensions $m_X = \int d\mu_X$ and $m_Y = \int d\mu_Y$ on $\mathcal{E}[\mathcal{A}_X] \otimes \mathcal{E}[\mathcal{A}_Y]$. This ring of functions contains the (indicator functions of the) rectangles and thus all of $\mathcal{A}_X \otimes \mathcal{A}_Y$. In fact, by its very definition it is the linear span of the rectangles. The restriction of $m_X \times m_Y$ to $\mathcal{A}_X \otimes \mathcal{A}_Y$ evidently agrees with $\mu_X \times \mu_Y$ on rectangles. It is the sought–after additive extension. The linear extension of any other additive set function satisfying $(*)$ agrees with $m_X \times m_Y$ on $\mathcal{E}[\mathcal{A}_X] \otimes \mathcal{E}[\mathcal{A}_Y]$ and therefore on $\mathcal{A}_X \otimes \mathcal{A}_Y$. In summary:

$$m_X \times m_Y = \int d(\mu_X \times \mu_Y) \;\; \text{on} \;\; \mathcal{E}[\mathcal{A}_X \otimes \mathcal{A}_Y] = \mathcal{E}[\mathcal{A}_X] \otimes \mathcal{E}[\mathcal{A}_Y] .$$

Example 1.2 (Product of Radon Measures) Suppose X, Y are locally compact spaces and m_X, m_Y are positive Radon measures. The ring $\mathcal{E} = C_{00}[X] \otimes C_{00}[Y]$ consists of continuous functions with compact support, but does not in general exhaust $C_{00}(X \times Y)$. Nevertheless, $\mathcal{E}$ is sufficiently dense in $C_{00}[X \times Y]$ so that there is a unique extension of $m_X \times m_Y$ to a positive Radon measure on $X \times Y$. This extension is called the product of the Radon measures. We leave the details as an exercise.

Exercise 1.3 Suppose $\mathcal{E}_X, \mathcal{E}_Y$ are lattice rings. Then every function $\phi \in \mathcal{E}$ can be written as the difference of two positive functions $\phi^+, \phi^- \in \mathcal{E}$. There exists a positive function $\psi \in \mathcal{E}$ with $|\phi| \le \psi$.

We assume henceforth that both m_X and m_Y are positive and σ–additive. We can then apply Daniell's procedure. The integrable functions we obtain will be called the m_X– or m_Y–integrable functions and their collections will be denoted by $\mathcal{L}^1[m_X]$ or $\mathcal{L}^1[m_Y]$, respectively. Clearly $m_X \times m_Y$ is positive as well. It is also σ–additive. However, we shall not prove this directly, but rather jump the gun and ask for a mean majorizing it — eventually we want to find one anyway. Consider

$$\int_{X \times Y}^{\flat} f = \int_X^* \left[\int_Y^* f(x,y)\, m_Y(dy) \right] m_X(dx)$$

and

$$\|f\|^{\flat} = \int_{X \times Y}^{\flat} |f| \,.$$

$\int_{X \times Y}^{\flat}$ is nothing but an iterated upper integral.

Lemma 1.4 *$\| \ \|^{\flat}$ is a mean that majorizes $m = m_X \times m_Y$ and that agrees on $(\mathcal{E}_X \otimes \mathcal{E}_Y)_+$ with m.*

Proof. The last statement first:

$$\left| \int \phi \, dm \right| \le \int_X \left[\int_Y |\phi(x,y)|\, m_Y(dy) \right] m_X(dx) = \|f\|^{\flat} \,,$$

with equality if $\phi \in \mathcal{E}^X \otimes \mathcal{E}^Y$ is positive. Note here also that $\| \ \|^{\flat}$ is finite on $\mathcal{E}_X \otimes \mathcal{E}_Y$: If ϕ is as in Equation (1.1), then

$$|\phi| \le \sum_{i=1}^{I} \left| \phi_X^{(i)} \right| \cdot \left| \phi_Y^{(i)} \right| \quad \text{and} \quad \| |\phi| \|^{\flat} \le \sum_{i=1}^{I} \int \left| \phi_X^{(i)} \right| dm_X \cdot \int \left| \phi_Y^{(i)} \right| dm_Y < \infty \,.$$

The solidity and absolute–homogeneity of $\| \ \|^{\flat}$ are evident, the finite subadditivity nearly so:

$$\begin{aligned} \|f+g\|^{\flat} &= \int_X^* \left[\int_Y^* |f+g| \, dm_Y \right] dm_X \le \int_X^* \left[\int_Y^* |f| + |g| \, dm_Y \right] dm_X \\ &\le \int_X^* \left[\int_Y^* |f| \, dm_Y \right] dm_X + \int_X^* \left[\int_Y^* |g| \, dm_Y \right] dm_X \\ &= \|f\|^{\flat} + \|g\|^{\flat} \,. \end{aligned}$$

The countable subadditivity follows from this and the continuity along arbitrary increasing sequences as in the proof of Theorem IV.1.6 (See also Exercise II.3.11 on p. 47); the latter is again easy to establish: Let (f_n) be an arbitrary increasing sequence of positive functions on $X \times Y$ with pointwise supremum f. Then $\left(\int_Y^* f_n(x,y)\, m_Y(dy)\right)$ increases to $\int_Y^* f(x,y)\, m_Y(dy)$ for every $x \in X$, and consequently

$$\|f_n\|^\flat = \int_X^* \left[\int_Y^* f_n(x,y)\, m_Y(dy)\right] m_X(dx)$$

increases to

$$\int_X^* \left[\int_Y^* f\,(x,y)\; m_Y(dy)\right] m_X(dx) = \|f\|^\flat\,.$$

It is left to be shown that $\|\;\|^\flat$ satisfies (M). This follows as in the proof of Theorem II.3.5 on p. 46 simply from the fact that $\|\;\|^\flat$ is additive on $\mathcal{E}_+$. ∎

Now that we have the mean $\|\;\|^\flat$ to majorize the elementary integral m it is evident what to do — or so it seems: we close $\mathcal{E}$ with respect to it to obtain $\mathcal{L}^1[\|\;\|^\flat]$, and we extend the elementary integral by continuity to obtain the extension $\int dm : \mathcal{L}^1[\|\;\|^\flat] \to \mathbb{R}$. We get the Dominated Convergence Theorem and all its beautiful consequences, simply because $\|\;\|^\flat$ is a mean. (By the way, it is for this situation that we developed the integration theory of means for rings in Section II.5 ff.: $\mathcal{E}_X \otimes \mathcal{E}_Y$ is a ring but not, in general, a lattice.)

We can evaluate the integral by iterated integration:

Proposition 1.5 *Let $f : X \times Y \to \mathbb{R}$ be a $\|\;\|^\flat$-integrable function. For m_X-almost all $x \in X$, the function $y \mapsto f(x,y)$ is m_Y-integrable. The integral*

$$x \mapsto \int f(x,y)\, m_Y(dy)$$

is thus m_X-almost everywhere defined. It is an m_X-integrable function of $x \in X$, and its m_X-integral is $\int f\, dm$:

$$\int f(x,y)\, d(m_X \times m_Y) = \int \left[\int f(x,y)\, m_Y(dy)\right] m_X(dx)\,.$$

Proof. (i): If f is $\|\;\|^\flat$-negligible, then the function $x \mapsto \int^* |f(x,y)|\, m_Y(dy)$ is m_X-negligible. In other words, *if f is $\|\;\|^\flat$-negligible, then $y \mapsto f(x,y)$ is m_Y-negligible for m_X-almost all $x \in Y$.*
(ii): Now suppose f is $\|\;\|^\flat$-integrable. Then there is a sequence (ϕ_n) in $\mathcal{E}$ such that

$$\sum_{n=1}^\infty \|\phi_n\|^\flat \le \infty \quad \text{and} \quad f(x,y) = \sum_{n=1}^\infty \phi_n(x,y) \quad \|\;\|^\flat\text{-almost everywhere.}$$

Then

$$\mathcal{N} \stackrel{\text{def}}{=} \left\{ (x,y) : \sum_{n=1}^{\infty} |\phi_n(x,y)| = \infty \quad \text{or} \quad f(x,y) \neq \sum_{n=1}^{\infty} \phi_n(x,y) \right\}$$

is a $\| \ \|^\flat$–negligible set. By (i)

$$\mathcal{N}_X^1 = \left\{ x \in X : \int_X^* \mathcal{N}(x,y)\, m_Y(dy) > 0 \right\}$$

is m_X–negligible. So is the set

$$\mathcal{N}_X^2 = \left\{ x \in X : \int^* \sum_n \big|\phi_n(x,y)\big|\, m_Y(dy) = \infty \right\} .$$

Consider an $x \in X$ that lies outside the m_X–negligible set $\mathcal{N}_X = \mathcal{N}_X^1 \cup \mathcal{N}_X^2$. Then $y \mapsto \sum |\phi_n(x,y)|$ has finite m_Y–upper integral and dominates the partial sums $y \mapsto \sum_{n=1}^{N} \phi_n(x,y)$ of elementary functions in $\mathcal{E}_Y$. The latter converge m_Y–almost everywhere to $y \mapsto f(x,y)$. The function $y \mapsto f(x,y)$ is therefore m_Y–integrable, and the integral is the limit of the numbers

$$I_N(x) = \int \sum_{n=1}^{N} \phi_n(x,y)\, m_Y(dy) , \qquad x \notin \mathcal{N}_X.$$

I_N is an elementary function in $\mathcal{E}_X$, majorized m_X–almost everywhere, namely in every point outside $\mathcal{N}_X$, by the function

$$x \mapsto \int^* \sum_{n=1}^{\infty} \big|\phi_n(x,y)\big|\, m_Y(dy) ,$$

whose m_X–upper integral, being majorized by $\left\| \sum |\phi_n| \right\|^\flat$, is finite. The pointwise limit $\int f(x,y)\, m_Y(dy)$ of the I_N is therefore m_X–integrable, and its integral equals

$$\begin{aligned}
\int \left[\int f(x,y)\, m_Y(dy) \right] m_X(dx) &= \lim_{N\to\infty} \int I_N(x)\, m_X(dx) \\
&= \lim_{N\to\infty} \int \left[\int \sum_{n=1}^{N} \phi_n(x,y)\, m_Y(dy) \right] m_X(dx) \\
&= \lim_{N\to\infty} \int \sum_{n=1}^{N} \phi_n \, dm = \int f(x,y)\, m(dx,dy) . \quad ∎
\end{aligned}$$

V.2 The Theorems of Fubini and Tonelli

These are the most powerful interchange–of–limits theorem in analysis. The setting is that of the previous section.

There is something unsatisfactory about the arguments of Section 1. Namely, everything depended on the choice of the order of integration. Had we defined he mean $\| \|^\flat$ by upper integration the other way around, first over x and then over y, we should probably have gotten a different mean and a different space $\mathcal{L}^1[\| \|^\flat]$ of integrable functions. Which one is the right one? Evidently neither; rather, we should define the extension by employing a mean that does not refer to the order of integration, indeed, not even to the fact that m comes as a product. The Daniell mean comes to mind; but the proof that Daniell's up–and–down–procedure actually produces a mean uses rather heavily the lattice structure of the elementary integrands, which we do not have at our disposal. At this point we remember the characterization III.6.2 (i) of Daniell's mean as the maximal mean for the elementary integral. *The natural choice of a mean for $m = m_X \times m_Y$ is the largest mean that agrees with m on $\mathcal{E}_+$.* This exists: the collection of means agreeing with m on $\mathcal{E}_+$ is not void, since it contains $\| \|^\flat$ (and the mean obtained by upper integration first over x and then over y), and the supremum of this collection is a mean, evidently the largest (see page 93). We denote this mean by $\| \|^*$ and call it the Daniell mean for m. We define $\mathcal{L}^1[m] = \mathcal{L}^1[m_X \times m_Y]$ to be the closure of $\mathcal{E}_X \otimes \mathcal{E}_Y$ in $\mathcal{F}[\| \|^*]$, and extend $m_X \times m_Y$ to $\mathcal{L}^1[m]$ by continuity. The functions in $\mathcal{L}^1[m] = \mathcal{L}^1[\| \|^*]$ are called ***$m_X \times m_Y$-integrable*** or ***integrable for the product measure***. The extensions of the integral are traditionally denoted by the same symbols as the elementary integrals, $\int_X$, $\int_Y$, and $\int d(m_X \times m_Y)$, respectively.

We know from the general theory that $\mathcal{L}^1[m]$ is a vector lattice closed under chopping, that the Dominated Convergence Theorem holds, what the relation of the integrable and measurable functions are, etc. The question remaining is whether we can evaluate $\int f(x,y)\, d(m_X \times m_Y)$ by iterated integration. Fubini's theorem says we can.

Theorem 2.1 (Fubini's Theorem) *Let $f : X \times Y \to \mathbb{R}$ be a function integrable for the product measure. (i) For m_X–almost every $x \in X$, the function $y \mapsto f(x,y)$ is m_Y–integrable. The integral*

$$x \mapsto \int f(x,y)\, m_Y(dy)$$

is thus m_X–almost everywhere defined.It is an m_X–integrable function of $x \in X$, and its m_X–integral is $\int f\, dm$:

$$\int f(x,y)\, d(m_X \times m_Y) = \int \left[\int f(x,y)\, m_Y(dy) \right] m_X(dx)\,.$$

(ii) The function $x \mapsto f(x,y)$ is m_X–integrable for m_Y–almost all y, and

$$\int f(x,y)\, d(m_X \times m_Y) = \int \left[\int f(x,y)\, m_X(dx)\right] m_Y(dy) .$$

Proof. (i): Since $\|\ \|^\flat \leq \|\ \|^*$, f is $\|\ \|^\flat$–integrable, and Proposition 1.5 on p. 128 yields the claim. (ii): interchange the rôles of X and Y. ▮

Theorem 2.1 is a powerful result. Yet it has a glaring shortcoming: in order to apply it to a function f one has to know in advance that f is integrable for the product measure. It would be much nicer if the integrability of f could also be decided by iterated (upper) integration. This is possible in a large number of cases:

Theorem 2.2 (Fubini–Tonelli) *Suppose $f : X\times Y \to \mathbb{R}$ is measurable for the product measure and has σ–finite carrier. Then f is integrable for the product measure if and only if one of the iterated upper integrals*

$$\int^* \left[\int^* |f(x,y)|\, m_Y(dy)\right] m_X(dx) \quad \textit{or} \quad \int^* \left[\int^* |f(x,y)|\, m_X(dx)\right] m_Y(dy)$$

is finite. In this case the iterated upper integrals agree and equal $\|f\|^*$, and

$$\begin{aligned}\int f(x,y)\, m_X \times m_Y(dx,dy) &= \int \left[\int f(x,y)\, m_Y(dy)\right] m_X(dx) \\ &= \int \left[\int f(x,y)\, m_X(dx)\right] m_Y(dy) .\end{aligned}$$

The merit of this result is this: very frequently both $(\mathcal{E}_X, m_X)$ and $(\mathcal{E}_Y, m_Y)$ will be σ–finite; then every function on $X\times Y$ will have σ–finite carrier. Just as frequently, the function f will also be given in terms that make it obvious that it is measurable, in view of the splendid permanence properties of that notion. In such a case only one of the iterated upper integrals will have to be checked for finiteness.

Proof. Again $\|\ \|^*$ is the Daniell mean for the product measure and $\|\ \|^\flat$ the mean obtained by iterated upper integration in either order. First assume f vanishes off some $\|\ \|^*$–integrable set A. Construct by induction a sequence (A^k) of mutually disjoint $\|\ \|^*$–integrable subsets of A with

$$\|A - (A^1 \cup A^2 \cup \ldots \cup A^k)\|^* < 2^{-k}$$

on which f is the uniform limit of elementary functions. Then $g^k = |f| \cdot A^1 + \cdots + |f| \cdot A^k$ is $\| \ \|^*$–integrable and so $\|g^k\|^* = \|g^k\|^\flat$. The g^k increase $\| \ \|^*$–almost everywhere to $|f| = |f| \cdot A$. The continuity along increasing sequences of $\| \ \|^*$ implies

$$\|f\|^* = \|f\|^\flat .$$

Next we cover the carrier of f with countably many $\| \ \|^*$–integrable sets A_n, choosing them to be increasing with n. By the continuity along increasing sequences of the means we have

$$\|f\|^* = \lim_n \|f \cdot A_n\|^* = \lim_n \|f \cdot A_n\|^\flat = \|f\|^\flat .$$

The same argument applies if $\| \ \|^\flat$ is defined by iterated upper integration the other way around. Thus if $\|f\|^\flat$ is finite then so is $\|f\|^*$, and as f is assumed to be measurable it is integrable. The theorem of Fubini does the rest. ∎

Corollary 2.3 *Let $f^X : X \to \mathbb{R}$ and $f^Y : Y \to \mathbb{R}$ be functions.*
If f^X is m_X–negligible (m_X–integrable, m_X–measurable) and f^Y is m_Y–negligible (m_Y–integrable, m_Y–measurable), then $f^X \cdot f^Y : X \times Y \to \mathbb{R}$ is negligible (integrable, measurable) for the product measure.

Proof. Suppose f^X, f^Y are negligible. Let $0 < \epsilon \leq 1$. There are $h^X \in \mathcal{E}_X^\uparrow$ and $h^Y \in \mathcal{E}_Y^\uparrow$ such that $\int h^X \, dm_X < \epsilon$, $\int h^Y \, dm_Y < \epsilon$ and $|f^X| \leq h^X$, $|f^Y| \leq h^Y$. Then $h^X \cdot h^Y$ belongs to $(\mathcal{E}_X \otimes \mathcal{E}_Y)^\uparrow$, and by Theorem 2.2 $\|h^X \cdot h^Y\|^* < \epsilon^2 \leq \epsilon$. We conclude that $f^X \cdot f^Y$ is negligible for the product measure.
Suppose next f^X, f^Y are integrable. There are sequences $(\phi_n^X), (\phi_n^Y)$ of elementary functions converging almost everywhere to f^X, f^Y. By the first part of the proof, $(\phi_n^X \cdot \phi_n^Y)$ converges almost everywhere to $f^X \cdot f^Y$, so that this function is measurable for the product measure. We may compute $\|f^X \cdot f^Y\|^*$ by iterated upper integrals (Theorem 2.2), finding that this number is finite. Thus $f^X \cdot f^Y$ is integrable for the product measure.
Lastly, suppose f^X, f^Y are measurable for the product measure. Their product is then clearly measurable on rectangles of the form $A^\diamond = A^X \times A^Y$, $A^X \subset X$ m_X–integrable, $A^Y \subset Y$ m_Y–integrable. These rectangles form an adequate cover of $X \times Y$. The details are left as an exercise. ∎

Here is a much–used consequence of the last theorem. It has many applications, notable to the estimation of p–means and other measures of the size of a function. The reader can find some of them in Section 4 and in the supplements to Section 10.

Corollary 2.4 *Let $(S, \mathcal{E}, \int d\mu)$ be a positive σ–additive elementary integral and f a positive μ–integrable function. Let $\Phi : [0, \infty) \to [0, \infty)$ be a continuous increasing* *function with* $\Phi(0)=0$ *and assume* Φ *has a continuous derivative* Φ' *on* $(0,\infty)$. *Then*

$$\int \Phi(f) = \int \Phi'(t)\cdot \mu([f>t])\,dt\,.$$

Proof. As usual we have written $\mu([f>t])$ for $\int_{[f>t]} d\mu$. The function $(x,t)\mapsto f(x)-t$ is measurable for the product measure $\mu\otimes\lambda$ on $\mathcal{E}\otimes\mathcal{E}[\mathbb{R}]$ (why?), and then so is $(x,t)\mapsto \Phi'(t)\cdot[f(x)>t]$. Taking the integral with respect to $\lambda(dt)$ first yields $\Phi(f)$, by the Fundamental Theorem of Calculus. Since the μ–upper integral of this is finite, $(x,t)\mapsto \Phi'(t)\cdot[f(x)>t]$ is integrable on the product space, and taking the integral first with respect to μ and then with respect to λ yields the same number. That is the formula to be proved. ∎

Exercise 2.5 (The Product of Order–continuous integrals) Suppose both m_X and m_Y are order–continuous. In the arguments of this and the last section replace $\|\ \|^\flat$ by

$$f\mapsto \int_X^\bullet \left[\int_Y^\bullet |f(x,y)|\,m_Y(dy)\right] m_X(dx)$$

and $\|\ \|^*$ by the largest order–continuous mean $\|\ \|^\bullet$ that agrees with $\|\ \|^\flat$ on $\mathcal{E}_+$ (Exercise III.6.5 on p. 94). Show that such actually exists and prove the theorems of Fubini and Tonelli–Fubini for $\|\ \|^\bullet$–integrable functions.

V.3 An Application: Convolution

Let f,g be Lebesgue integrable functions on the line. Their *convolution* is the function $f*g$ given by

$$f*g(y)=\int_{\mathbb{R}} f(x)g(y-x)\,\lambda(dx)\,. \tag{3.1}$$

Proposition 3.1 *The integral (3.1) exists for* λ*–almost all* $y\in\mathbb{R}$. *The function* $f*g$ *is integrable.*

Proof. This would follow rather easily from Theorem 2.2 if $(x,y)\mapsto g(y-x)$ were known to be measurable for the product measure $\lambda(dx)\times\lambda(dy)$. It isn't, so we have to approach (3.1) obliquely. Let $\underline{f}\le f\le\overline{f}$, $\underline{g}\le g\le\overline{g}$, etc. be the *Baire measurable* upper and lower envelopes provided by Proposition III.6.2. Observe that the absolute value of the function

$$\begin{aligned}\Delta(x,y)&=f(x)g(y-x)-\underline{f}(x)\underline{g}(y-x)\\&=\Big(f(x)-\underline{f}(x)\Big)\cdot g(y-x)+\underline{f}(x)\cdot\Big(g(y-x)-\underline{g}(y-x)\Big)\end{aligned}$$

is majorized by

$$\left|\overline{f}(x) - \underline{f}(x)\right| \cdot \overline{|g|}(y-x) + \overline{|f|}(x) \cdot \left|\overline{g}(y-x) - \underline{g}(y-x)\right| .$$

The two summands on the right are Baire measurable and negligible for $\lambda(dx)\times\lambda(dy)$. Indeed, to see that the second summand, for example, is negligible we may integrate first over y and then over x, by Theorem 2.2. The integration over y gives

$$\overline{|f|}(x) \cdot \int |\overline{g}(y-x) - \underline{g}(y-x)|\, dy = \overline{|f|} \cdot \int |\overline{g}(y) - \underline{g}(y)|\, dy = 0$$

by the translation–invariance of Lebesgue measure. We see that Δ is $\lambda(dx)\times\lambda(dy)$–negligible. Consider now the function

$$f(x)g(y-x) - \Delta(x,y) = \underline{f}(x)\underline{g}(y-x) . \tag{3.2}$$

$y-x$ is a Baire function on the plane. So is $\underline{g}(y-x)$, since the Baire functions are closed under left–composition with Baire functions on the line (Proposition III.5.16 on p. 91). The function of (3.2) is thus a Baire function on the plane. Integration of the absolute value of this function over y gives $\int |g(y)|\,\lambda(dy)\cdot|\underline{f}(x)|$; integration over x gives $\int |g(y)|\,\lambda(dy) \cdot \int |f(x)|\,\lambda(dx) < \infty$: the function of (3.2) is integrable for $\lambda(dx)\times\lambda(dy)$. Then so is $f(x)g(y-x)$, and Fubini's theorem yields the claim.

∎

Exercise 3.2 Let $f, g, h \in \mathcal{L}^1[\lambda]$. Then $f * g = g * f$ and $(f*g)*h = f*(g*h)$. The Fourier transforms satisfy $\mathcal{F}[f*g] = \mathcal{F}f \cdot \mathcal{F}g$.

Exercise 3.3 Let $m = f\lambda$ and $n = g\lambda$ be the measures with density f, g, respectively. Then the measure with density $f*g$ is given by

$$\int \phi(z)\,((f*g)\lambda)(dz) = \int \phi(x+y) f(x) g(y)\,\lambda(dx)\lambda(dy)$$

Exercise 3.4 $f*g$ is continuous.

V.4 An Application: Marcinkiewicz Interpolation

Interpolation is a large field. We prove here one rather simple result. The main idea is to introduce the reader to some interesting notions, that have proved to be very powerful. The results of this section will not be used later on.

Lorentz spaces. Let $(\mathcal{E}, m)$ be a positive σ–additive integral, and $1 \le p \le \infty$. Functions f, g etc. that should be measurable in the arguments below are understood to be so. Chebyshev's inequality (Exercise II.3.7) says that $\alpha^p \cdot m[|f| > \alpha] \le \int |f|^p \, dm$ and can be read as

$$\|f\|_{p,\infty} \stackrel{\text{def}}{=} \sup_{\alpha>0} \left\{ \alpha \cdot \Big(m[|f| > \alpha] \Big)^{1/p} \right\} \le \|f\|_p \, .$$

The number $\|f\|_{p,\infty}$ is thus a somewhat less stringent gauge for the size of f than $\|f\|_p$. For $p = \infty$ we set $\|f\|_{\infty,\infty} = \|f\|_\infty$. Except for this last one, the $\| \ \|_{p,\infty}$ are not subadditive. The spaces of functions they describe are not normable. It is frequently rather easier to show that a given function belongs to the ***Lorentz space***

$$\mathcal{L}^{p,\infty} = \Big\{ f \ m\text{–measurable} : \ \|f\|_{p,\infty} < \infty \Big\}$$

than to $\mathcal{L}^p$. For instance, in Section 10, which deals with differentiation, we shall meet the Hardy–Littlewood Maximal Operator. It is an operator of great utility in many contexts. It does not map $\mathcal{L}^1[\lambda]$ into itself but rather only into $\mathcal{L}^{1,\infty}[\lambda]$.

Exercise 4.1 The spaces $\mathcal{L}^{p,\infty}$ are complete metrizable topological vector spaces.

Operators of Weak Type. Let $(\mathcal{F}, n)$ be a second positive σ–additive integral, and consider a map U that associates with every elementary function $\phi \in \mathcal{F}$ an m–measurable function $U(\phi)$. Such U is ***of weak type p-q*** if there is a constant A such that for all $\phi \in \mathcal{F}$

$$\|U(\phi)\|_{p,\infty} \le A \cdot \|\phi\|_{\mathcal{L}^q(n)} \, .$$

It is of ***strong type p–q*** if instead

$$\|U(\phi)\|_{\mathcal{L}^p(m)} \le A \cdot \|\phi\|_{\mathcal{L}^q(n)} \, ,$$

in other words if it is continuous at zero, from $\mathcal{L}^q(n)$ to $\mathcal{L}^p(m)$. Here both p and q are in $[1, \infty]$. An operator of strong type p–q is clearly of weak type p–q; for $p = \infty$ the two notions coincide.

Exercise 4.2 If U is of weak type p–q then it has a unique extension to all of $\mathcal{L}^q(n)$ that is still of weak type p–q.

As an instructive application of Corollary 2.4 on p. 132 let us prove

Theorem 4.3 (An Interpolation Theorem) *Suppose $U : \mathcal{F} \to \mathcal{L}^0(m)$ is a subadditive operator of weak types p_1–p_1 and p_2–p_2 with constants A'_{p_1}, A'_{p_2}, respectively. Then U is of strong type p–p for $p_1 < p < p_2$:*

$$\|U(f)\|_{\mathcal{L}^p(m)} \le A_p \cdot \|f\|_{\mathcal{L}^q(n)}$$

with constant

$$A_p \le p^{1/p} \cdot \Big(\frac{(2A'_{p_1})^{p_1}}{p - p_1} + \frac{(2A'_{p_2})^{p_2}}{p_2 - p} \Big)^{1/p} .$$

Proof. By the subadditivity of U we have for every $t>0$

$$|U(f)| \le |U(f\cdot[|f|\ge t])| + |U(f\cdot[|f|<t])|$$

and consequently

$$\begin{aligned} m[|U(f)|\ge t] &\le m[|U(f\cdot[|f|\ge t])|\ge t/2] + m[|U(f\cdot[|f|<t])|\ge t/2] \\ &\le \left(\frac{A'_{p_1}}{t/2}\right)^{p_1}\int_{[|f|\ge t]}|f|^{p_1}\,dn + \left(\frac{A'_{p_2}}{t/2}\right)^{p_2}\int_{[|f|<t]}|f|^{p_2}\,dn\,. \end{aligned}$$

Multiply with t^{p-1} and integrate against dt, using Fubini's theorem:

$$\begin{aligned} \int |U(f)|^p\,dm\Big/p &= \int\int_0^{|U(f)|} t^{p-1}\,dtdm = \int\int_0^\infty m[|U(f)|\ge t]t^{p-1}\,dt \\ &\le \int\left(\frac{A'_{p_1}}{t/2}\right)^{p_1}\int_{[|f|\ge t]}|f|^{p_1}t^{p-1}\,dtdn \\ &\quad + \int\left(\frac{A'_{p_2}}{t/2}\right)^{p_2}\int_{[|f|<t]}|f|^{p_2}t^{p-1}\,dtdn \\ &= (2A'_{p_1})^{p_1}\int\int_{[|f|\ge t]}|f|^{p_1}t^{p-p_1-1}\,dtdn \\ &\quad + (2A'_{p_2})^{p_2}\int\int_{[|f|<t]}|f|^{p_2}t^{p-p_2-1}\,dtdn \\ &= (2A'_{p_1})^{p_1}\int\int_0^{|f|}|f|^{p_1}t^{p-p_1-1}\,dtdn \\ &\quad + (2A'_{p_2})^{p_2}\int\int_{|f|}^\infty|f|^{p_2}t^{p-p_2-1}\,dtdn \\ &= \frac{(2A'_{p_1})^{p_1}}{p-p_1}\int|f|^{p_1}\cdot|f|^{p-p_1}\,dn \\ &\quad + \frac{(2A'_{p_2})^{p_2}}{p_2-p}\int|f|^{p_2}\cdot|f|^{p-p_2}\,dn \\ &= \left(\frac{(2A'_{p_1})^{p_1}}{p-p_1}+\frac{(2A'_{p_2})^{p_2}}{p_2-p}\right)\cdot\int|f|^p\,dn\,. \end{aligned}$$

Multiply both sides with p and take p^{th} roots; the claim follows. ∎

Note that the constant A_p blows up as $p\downarrow p_1$ or $p\uparrow p_2$. Here is a situation in which more can be said:

Exercise 4.4 Suppose that $(\mathcal{E}, m) = (\mathcal{F}, n)$ and that the map U happens to be ***self–adjoint***. This means that

$$\int U(\phi) \cdot \psi \, dm = \int \phi \cdot U(\psi) \, dm \qquad \phi, \psi \in \mathcal{E} .$$

If U of weak types 1–1 and 2–2 then U is of strong type p–p for all $p \in (1, \infty)$.

V.5 Signed Measures

There do occur linear maps $\int = m : \mathcal{E} \to \mathbb{R}$ on lattice rings that are not positive. They are called ***signed elementary integrals*** if attention is to be drawn to the non–positivity. For instance, the continuous linear functionals on the space $C[K]$, K compact, form a large class of signed elementary integrals that is natural to investigate. For another example imagine a physicist dealing with plasma. He may wish to describe the charge $\mu(dV)$ in a volume dV. If dV contains mostly electrons then $\mu(dV)$ should be negative, if protons predominate it should be positive. He faces an additive set function μ that takes both positive and negative values, a ***signed measure***. To reduce the argument to a common denominator we have the physicist extend his measure immediately by linearity to the step functions, as in Proposition II.2.13 on p. 39. He obtains a signed elementary integral m.

Naturally, we would like to extend $\int \bullet \, dm$ so as to obtain as powerful a calculus as in the case of positive measures. The following idea leaps to the mind: let us find a mean $\| \ \|^*$ that majorizes $\int$, and use it to extend $\int$ by continuity to the $\| \ \|^*$–closure $\mathcal{L}^1$ of $\mathcal{E}$. Since the permanence properties of the extended integral were all due to the majorizing mean, we should then have them in the present situation as well.

This idea succeeds. The elementary integral must, of course, meet some conditions in order to admit a majorizing mean, one more than in the positive case.

First, it must be σ–additive, or σ–continuous, which by Exercise II.2.2 on p. 36 is the same thing. To see this let (ϕ_n) be an increasing sequence of elementary integrands whose pointwise supremum ϕ belongs again to $\mathcal{E}$. Since $\| \ \|^*$ is a mean the Monotone Convergence Theorem holds, and $\phi_n \to \phi$ in mean. Since the elementary integral is continuous with respect to the mean, $\int \phi_n \to \int \phi$.

Second, $\int$ must have finite variation. This condition replaces the positivity. To see what this means, fix a $\psi \in \mathcal{E}_+$. The solidity of the prospective majorizing mean implies that for any elementary integrand ϕ with $|\phi| \le \psi$

$$\left| \int \phi \right| \le \left\| |\phi| \right\|^* \le \|\psi\|^* .$$

Taking the supremum over $\phi \in \mathcal{E}$ with $|\phi| \le \psi$ we see that $\int$ must meet the

Definition 5.1 *The elementary integral m has* ***finite variation*** *if the number*

$$\left|m\right|(\psi) = \sup\left\{\left|\int \phi\, dm\right| : \phi \in \mathcal{E}\,, |\phi| \le \psi\right\} \tag{5.1}$$

is finite for every $\psi \in \mathcal{E}_+$. $|m|(\psi)$ is called the ***variation*** *of m on ψ and is also denoted by $\int d|m|$.*

This notation anticipates point (ii) of the following result:

Lemma 5.2 *(i) For any positive $\psi \in \mathcal{E}$*

$$|m|(\psi) = \sup\{m(\phi_1) - m(\phi_2) : \phi_i \in \mathcal{E}_+, \phi_1 + \phi_2 = \psi\}\,. \tag{5.2}$$

(ii) Suppose the elementary integral $(\mathcal{E}, m)$ has finite variation. Then the map $\psi \mapsto |m|(\psi)$ is positive–homogeneous, increasing, and additive on $\mathcal{E}_+$. There exists a unique extension to a positive elementary integral, which will again be denoted by $|m|$. It is the smallest positive elementary integral majorizing both m and $-m$ and satisfies

$$|m(\phi)| \le |m|(|\phi|)\,, \quad \forall\, \phi \in \mathcal{E}\,. \tag{5.3}$$

Proof. (i): Suppose the right–hand side of Equation (5.1) strictly exceeds r. Then there is a $\phi \in \mathcal{E}$ with $|\phi| \le \psi$ and $m(\phi) > r$. We have $m(\phi) = m(\phi_+) - m(\phi_-)$, where $\phi_\pm \in \mathcal{E}_+$ and $\phi_+ + \phi_- \le \psi$. Let $\delta = \psi - (\phi_+ + \phi_-)$. If $m(\delta) > 0$ we set $\phi_1 = \phi_+ + \delta, \phi_2 = \phi_-$, else we set $\phi_1 = \phi_+, \phi_2 = \phi_- + \delta$. In either case we get $\phi_1 + \phi_2 = \psi$ and $m(\phi_1) - m(\phi_2) > r$: the right–hand side of (5.2) exceeds r as well and is therefore always greater than that of Equation (5.1). Conversely, assume the right–hand side of (5.2) strictly exceeds r. There are then $\phi_i \in \mathcal{E}_+$ with $\phi_1 + \phi_2 = \psi$ and $m(\phi_1) - m(\phi_2) > r$. If the second integral is positive set $\phi = \phi_1$, else set $\phi = \phi_1 - \phi_2$. Clearly $|\phi| \le \psi$ and $m(\phi) > r$: the right–hand side of Equation (5.1) exceeds r as well and is therefore always greater than that of (5.2). The two are the same.
(ii): The monotonicity and positive–homogeneity are left as an exercise. The additivity is best shown in two steps.
1) $|m|$ is subadditive: Let $\psi_1, \psi_2 \in \mathcal{E}_+$ and assume $|m|(\psi_1 + \psi_2) > r$. There exist $\phi^1, \phi^2 \in \mathcal{E}_+$ with sum $\psi_1 + \psi_2$ and $m(\phi^1) - m(\phi^2) > r$. Set

$$\begin{array}{rclrcl} \phi_1^1 & = & \psi_1 \wedge \phi^1\,, & \phi_1^2 & = & \phi^1 - \psi_1 \wedge \phi^1\,, \\ \phi_2^1 & = & \psi_1 - \psi_1 \wedge \phi^1\,, & \phi_2^2 & = & \psi_2 + \psi_1 \wedge \phi^1 - \phi^1\,. \end{array}$$

These four functions are positive elementary integrands. This is obvious for all but ϕ_2^2; and for this it follows from $\psi_2 + \psi_1 \wedge \phi^1 - \phi^1 = (\psi_1 + \psi_2 - \phi^1) \wedge \psi_2 = \phi^2 \wedge \psi_2$. The columns of this matrix add up to ψ_1 and ψ_2, the rows to ϕ^1 and ϕ^2. Evidently

$$\begin{aligned} |m|(\psi_1) + |m|(\psi_2) &\ge \left(m(\phi_1^1) - m(\phi_1^2)\right) + \left(m(\phi_2^1) - m(\phi_2^2)\right) \\ &= m(\phi_1^1 + \phi_2^1) - m(\phi_1^2 + \phi_2^2) = m(\phi^1) - m(\phi^2) > r\ : \end{aligned}$$

the subadditivity is established.

2) $¦m¦$ *is superadditive:* Let $\psi_1, \psi_2 \in \mathcal{E}_+$ with $¦m¦(\psi_1) + ¦m¦(\psi_2) > r$. Then there are $\phi_i^1, \phi_i^2 \in \mathcal{E}_+$ with $\psi_i = \phi_i^1 + \phi_i^2$ and

$$\Big(m(\phi_1^1) - m(\phi_1^2)\Big) + \Big(m(\phi_2^1) - m(\phi_2^2)\Big) > r\,.$$

This reads $m(\phi^1) - m(\phi^2) > r$, where $\phi^1 = \phi_1^1 + \phi_2^1$ and $\phi^2 = \phi_1^2 + \phi_2^2$ are positive elementary integrands with sum $\psi_1 + \psi_2$. We conclude that $¦m¦(\psi_1 + \psi_2) > r$: The superadditivity is established as well, and with it the additivity.
The additivity can also be shown along similar lines using the defining Equation (5.1). We leave this as an exercise.

We are ready to tackle the remaining claims. If $¦m¦$ is the restriction of a positive elementary integral on all of $\mathcal{E}$ then necessarily

$$¦m¦(\psi) = ¦m¦(\psi_1) - ¦m¦(\psi_2) \tag{$*$}$$

whenever $\psi \in \mathcal{E}$ is the difference of the positive functions ψ_1, ψ_2 in $\mathcal{E}_+$. Such an extension is thus unique. Conversely, since every $\psi \in \mathcal{E}$ is the difference of two positive functions we may use $(*)$ to define the extension of $¦m¦$ to all of $\mathcal{E}$. This definition is good; for if ψ is also written as the difference of $\psi_1', \psi_2' \in \mathcal{E}_+$, then $\psi_1 + \psi_2' = \psi_2 + \psi_1'$ and consequently $¦m¦(\psi_1) - ¦m¦(\psi_2) = ¦m¦(\psi_1') - ¦m¦(\psi_2')$. The last statement is nearly obvious from Equation (5.2): if n is a positive elementary integral majorizing both m and $-m$ then $n(\psi) \ge m(\phi_1) - m(\phi_2)$ whenever $\phi_i \in \mathcal{E}_+$ sum to $\psi \in \mathcal{E}_+$, and thus $n \ge ¦m¦$. The Inequality (5.3) is plain from the definition (5.1). ∎

Proposition 5.3 *The elementary integral m on $\mathcal{E}$ has finite variation if and only if there exists a positive elementary integral n majorizing both m and $-m$. In this case, $¦m¦$ is the smallest positive elementary integral majorizing both m and $-m$, and its value at a function $\psi \in \mathcal{E}_+$ is given by Equation (5.1) or (5.2).*

If the elementary integral m on $\mathcal{E}$ has finite variation and is σ–additive then its variation $¦m¦$ is also σ–additive.

Proof. The first statement is but a summary of Lemma 5.2. For the second, let ψ_n be an increasing sequence of elementary functions, without loss of generality positive, whose pointwise supremum is another elementary function ψ. It is to be shown that $\sup_n ¦m¦(\psi_n) \ge ¦m¦(\psi)$, the reverse inequality being obvious. Now if $¦m¦(\psi) > r$ then there is a $\phi \in \mathcal{E}$ with $|\phi| \le \psi$ and $|m(\phi)| > r$. Consider then the sequences $\phi_n^+ = \psi_n \wedge \phi_+$ and $\phi_n^- = \psi_n \wedge \phi_-$. They increase pointwise to ϕ_+ and ϕ_-, respectively. In consequence of the σ–additivity of m we have $|m(\phi_n^+ - \phi_n^-)| > r$ for sufficiently high index n. From $|\phi_n^+ - \phi_n^-| \le \psi_n$ we conclude $\sup_n ¦m¦(\psi_n) > r$ and $\sup_n ¦m¦(\psi_n) \ge ¦m¦(\psi)$. ∎

The Integration Theory of a signed σ–additive elementary integral of finite variation is now obvious. Namely, we simply let $\| \ \|^*$ be Daniell's mean for the positive σ–additive elementary integral $¦m¦$, define $\mathcal{L}^1[m] = \mathcal{L}^1[¦m¦]$ as the closure of $\mathcal{E}$ under this mean, and extend the integral by continuity. This is possible since by (5.3) $\| \ \|^*$ majorizes the elementary integral:

$$|m(\phi)| \le ¦m¦(|\phi|) \le \|\phi\|^* , \qquad \forall \phi \in \mathcal{E} .$$

This is all very satisfactory and simple. A few questions remain:
1) Which Radon measures have finite variation and are σ–additive?
2) Suppose μ is a measure on a ring $\mathcal{A}$ of sets. How can we determine by just looking at μ whether its linear extension $m = \int d\mu$ to the step functions $\mathcal{E}[\mathcal{A}]$ has finite variation and is σ–additive?
3) What is the structure of the set $\mathfrak{M}[\mathcal{E}]$ of all elementary integrals on $\mathcal{E}$?
We defer 3) to the next section and answer the first two questions now.

Example: Radon Measures

Let us address the first question raised above. Here we have to remedy an earlier omission. Namely, in Lemma II.2.4 on p. 37 we did define only the notion of a *positive* Radon measure. Let S be a locally compact space and $C_{00}[S]$ the lattice ring of bounded continuous functions of compact support. A ***Radon measure*** is a linear functional $m : C_{00}[S] \to \mathbb{R}$ that is continuous in the following sense: if (ϕ_n) is a sequence of functions in $C_{00}[S]$ whose supports are all contained in a common compact set and which converges uniformly to a function ϕ then $m(\phi_n) \xrightarrow[n\to\infty]{} m(\phi)$. If S happens to be compact this means that m is a continuous linear functional on the normed space $(C[S], \| \ \|_u)$.

Exercise 5.4 A positive Radon measure is a Radon measure.

Proposition 5.5 *A linear functional $m : C_{00}[S] \to \mathbb{R}$ is a Radon measure precisely if it has finite variation. It is then σ–additive, in fact order–continuous.*

Proof. If the variation of m is infinite one can find $\phi_n, \psi \in C_{00}[S]$ with $|\phi_n| \le \psi$ and $m(\phi_n) > 2^n$. The functions $2^{-n}\phi_n$ all vanish off the support of ψ and converge uniformly to zero. Since $|m(2^{-n}\phi_n)| > 1$, m is not a Radon measure. Conversely, if m is not Radon, then there is a sequence (ϕ_n) of functions in $C_{00}[S]$ whose supports are contained in a common compact set $K \subset S$, which converge uniformly to some $\phi \in C_{00}[S]$, and so that $|m(\phi_n - \phi)| > \epsilon$ for all n and some $\epsilon > 0$. There is a $\psi \in C_{00+}[S]$ that equals 1 on K. By extracting a subsequence if necessary we can arrange matters so that $\|\phi_n - \phi\|_u < 2^{-n}$. Consider $\psi_n = \pm(\phi_n - f)$, where the sign is chosen so that $m(\psi_n) > 0$. The functions $\Phi_N = \sum_{n=1}^N \psi_n$ all have $|\Phi_N| \le \psi$, and $m(\Phi_N) > N\epsilon$: m does not have finite variation. We conclude with Lemma II.2.4 on p. 37: since the variation $\|m\|$ of the Radon measure m is a positive Radon measure it is σ–additive, and therefore so is m itself. ∎

Example: Additive Set Functions of Finite Variation

We turn to the second question raised above. Let μ be a measure on a ring $\mathcal{A}$ and $m = \int d\mu$ its linear extension to the step functions $\mathcal{E}[\mathcal{A}]$. Suppose m does have finite variation $\|m\|$. Then the restriction $\|\mu\|$ of $\|m\|$ to $\mathcal{A}$ is clearly a positive measure majorizing both μ and $-\mu$. In fact, it is the smallest such measure. For if ν is another measure with this property then its linear extension $n = \int d\nu$ majorizes both m and $-m$ and thus exceeds $\|m\|$; we conclude that $\|\mu\| \leq \nu$. Conversely, assume that there is a positive measure ν on $\mathcal{A}$ that majorizes both μ and $-\mu$. Its linear extension then majorizes both $\int d\mu$ and $-\int d\mu$, so that $m = \int d\mu$ has finite variation. As before we see that there then exists a smallest positive measure exceeding both μ and $-\mu$, namely the restriction of $\|m\|$ to $\mathcal{A}$. It is therefore reasonable to call $\mu : \mathcal{A} \to \mathbb{R}$ a ***measure of finite variation*** if there exists a positive measure ν majorizing both μ and $-\mu$, and to denote the smallest such ν by $\|\mu\|$ and call it ***the variation of the measure μ***. The upshot of all this is listed as the first part of the following

Proposition 5.6 *Let μ be a measure on a ring $\mathcal{A}$ of sets.*
(i) The linear extension $\int d\mu$ of μ to the step functions over $\mathcal{A}$ has finite variation if and only if μ does. In this case, the variation of $\int d\mu$ is $\int d\|\mu\|$.
(ii) The variation $\|\mu\|$ at a set $A \in \mathcal{A}$ is given by

$$\|\mu\|(A) = \sup\{\mu(A_1) - \mu(A_2) : \ A_i \in \mathcal{A},\ A_1 + A_2 = A\} . \tag{5.4}$$

(iii) $\|\mu\|$, $m = \int d\mu$, and $\|m\| = \int d\|\mu\|$ are σ–additive if and only if μ is.

Proof. (ii): The proof of the additivity of $\|m\|$ in Lemma 5.2 on p. 138 applies literally, if the ϕ^i, ψ_j are taken to be sets of $\mathcal{A}$, and shows that the right–hand side of Equation (5.4) defines an additive measure. It is clearly the smallest positive measure majorizing both μ and $-\mu$. The details are left to the reader. ∎

Exercise 5.7 The map that associates with every measure of finite variation on a ring of sets its linear extension to the step functions over that ring is a linear order preserving bijection.

Exercise 5.8 The map of Exercise 5.7 maps the vector space of σ–additive measures of finite variation on the ring onto the vector space of σ–continuous elementary integrals of finite variation on the step functions.

Example: Distribution Functions on the Line

Let $F : \mathbb{R} \to \mathbb{R}$ be a function. F gives rise to a measure on $\mathcal{A}[\mathbb{R}]$, by the simple formula

$$\mu_F((a, b]) = F(b) - F(a) .$$

At first, this rule assigns a measure to intervals $(a, b]$ only. An arbitrary set A in $\mathcal{A}[\mathbb{R}]$ can be written as the union

$$A = \bigcup_{i=1}^{I} (a_i, b_i]$$

of *disjoint* intervals (Exercise I.4.6 on p. 20), and additivity requires that we set

$$\mu_F(A) = \sum_{i=1}^{I} F(b_i) - F(a_i) .$$

It is clear that $\mu_F : \mathcal{A}[\mathbb{R}] \to \mathbb{R}$ is an additive measure. If F and F' differ by a constant then the associated measures μ_F and $\mu_{F'}$ are the same.

Conversely, if μ is a measure on $\mathcal{A}[\mathbb{R}]$ then

$$F(t) = \begin{cases} \mu\Big((0, t]\Big) & \text{if } t \geq 0 \\ -\mu\Big((t, 0]\Big) & \text{if } t < 0 \end{cases}$$

defines a function F with $F(0) = 0$ whose associated measure is μ. F is called ***a distribution function*** of μ. We get a linear bijection between finite functions $F : \mathbb{R} \to \mathbb{R}$ that vanish at the origin, and measures on $\mathcal{A}[\mathbb{R}]$.

How can we tell from looking at the distribution function F whether the associated measure $\mu = \mu_F$ has finite variation? In answer to this question we establish a formula for the variation $\lvert\mu\rvert$ in terms of F: For any interval $(a, b]$,

$$\lvert\mu_F\rvert \Big((a, b]\Big) = \sup\Big\{ \sum_{i=1}^{I} |F(a_{i+1}) - F(a_i)| : \quad a = a_1 < a_2 < \cdots < a_{I+1} = b \Big\} . \tag{5.5}$$

The supremum is taken over all finite partitions $a = a_1 < a_2 < \cdots < a_{I+1} = b$ of $(a, b]$. If a function F is such that the right–hand side of (5.5) is finite for all $a < b \in \mathbb{R}$ then F is said to have ***finite variation***. This is evidently a most appropriate choice of words. (5.5) explains where the name finite variation for measures on the line — and then for arbitrary measures — comes from.

Proof of (5.5). If $\lvert\mu_F\rvert\big((a,b]\big) > r$ then there are, by Proposition 5.6 (ii), disjoint sets $A, B \in \mathcal{A}$ with union $(a,b]$ and $\mu_F(A) - \mu_F(B) > r$. We write A, B as unions of disjoint intervals $(a_i, b_i]$ and inspect: the right–hand side of (5.5) exceeds r. Conversely, if the right–hand side of (5.5) is strictly bigger than r then there is a partition $a = a_1 < a_2 < \cdots < a_{I+1} = b$ of $(a,b]$ with $\sum_{i=1}^{I} |F(a_{i+1}) - F(a_i)| > r$. We let A be the union of the intervals $(a_i, a_{i+1}]$ for which $F(a_{i+1}) - F(a_i)$ is positive, and B the union of the remaining partition intervals. Evidently $\mu_F(A) - \mu_F(B) > r$ and thus $\lvert\mu_F\rvert\big((a,b]\big) > r$. The equality is established. ∎

Suppose now F has finite variation. Let $F^{\pm}$ be distribution functions for the positive and negative parts of μ_F, respectively, and C a constant such that $F(0) = C + F^+(0) - F^-(0)$. Then $F = (C + F^+) - F^-$, because the measures associated with either side of the equality are the same; the equation exhibits F as the difference of two increasing functions: *A function of finite variation can be written as the difference of two increasing functions. It has therefore right and left limits at every point and is discontinuous at no more than countably many points.*

Assume again F is a function of finite variation. When is the associated measure σ–additive? The answer is simple:

Lemma 5.9 *μ_F is σ–additive if and only if F is right–continuous.*

Proof. If μ_F is σ–additive then, for every decreasing sequence (x_n) with limit x, $F(x_n) - F(x) = \mu_F\Big((x, x_n]\Big) \xrightarrow[n\to\infty]{} 0$, because $(x, x_n] \downarrow \emptyset$: F is right–continuous. Conversely, assume F is right–continuous. Write F as the difference of two increasing functions, $F = F^+ - F^-$. Define the ***right–continuous version*** of $F^{\pm}$ by $\underleftarrow{F}^{\pm}(x) = \lim_{y \downarrow x} F^{\pm}(y)$. Then $F = \underleftarrow{F} = \underleftarrow{F}^{+} - \underleftarrow{F}^{-}$, and we have succeeded in writing F as the difference of two increasing right–continuous functions. It suffices evidently to show that the measures associated with $\underleftarrow{F}^{\pm}$ are σ–additive. In other words, we may assume that F is increasing.

Consider then a decreasing sequence (D^n) of $\mathcal{A}[\mathbb{R}]$ with void intersection. It is to be shown that $\mu_F(D^n) \to 0$. To this end let $\epsilon > 0$ be given. Every D^n can be written as the union of finitely many elementary intervals:

$$D^n = \bigcup_{i=1}^{I(n)} (a_i^n, b_i^n] .$$

Because of the right–continuity of F there are points $b_i^m > a_i^m$ with $F(b_i^m) - F(a_i^m) < \epsilon 2^{-i-m}$. Set

$$\widehat{D}^n = D^n \Bigg\backslash \left(\bigcup_{m=1}^{n} \bigcup_{i=1}^{I(m)} (a_i^m, b_i^m) \right) .$$

The $\widehat{D}^n$ are compact and decrease to the void set. Thus one of them, and then all with higher index will be void. Say $\widehat{D}^N = \emptyset$. Then

$$D^n \subset \bigcup_{m=1}^{n} \bigcup_{i=1}^{I(m)} (a_i^m, b_i^m] \qquad \text{for } n \geq N .$$

Clearly, D^n has μ_F–measure less than $\sum_{m=1}^{\infty} \sum_{i=1}^{\infty} 2^{-i-m}\epsilon = \epsilon$. That is to say, $\mu_F(D^n) \xrightarrow[n\to\infty]{} 0$. This is but a copy of the proof of Lemma II.1.1, adapted to the present situation. ∎

V.6 The Space of Measures

We have now pretty exhaustively studied single elementary integrals on our lattice ring $\mathcal{E}$, have even looked at two of them simultaneously. The time has come to change our point of view and to study the totality of all elementary integrals on $\mathcal{E}$, "to explore the forest instead of its single trees," as it were. Accordingly let $\mathfrak{M}[\mathcal{E}]$ denote the set of all elementary integrals on $\mathcal{E}$ that have finite variation. This is a vector space. For if m, n have finite variation and $r, s \in \mathbb{R}$ then both $rm + sn$ and $-(rm + sn)$ are majorized by the positive elementary integral $|r|\,¦m¦ + |s|\,¦n¦$, and thus $rm + sn \in \mathfrak{M}[\mathcal{E}]$. This vector space has a natural order given by the order cone $\mathfrak{M}[\mathcal{E}]_+$ of positive elementary integrals: $m \leq n \iff n - m \geq 0$. It is plain that this order is compatible with the linear structure. We have seen in Lemma 5.2 (ii) that there is a least upper bound $¦m¦$ for m and $-m$. We could invoke Exercise IV.3.3 on p. 109 to conclude that $\mathfrak{M}[\mathcal{E}]$ is a vector lattice. Let us instead show directly that two elements m, n have a greatest lower bound $m \wedge n$. To this end let $\rho \in \mathfrak{M}[\mathcal{E}]$ and consider the following chain of implications:

$$\begin{aligned}
& m \geq \rho \ \text{ and } \ n \geq \rho \\
& \Longrightarrow n - m \leq n - \rho \ \text{ and } \ m - n \leq m - \rho \\
& \Longrightarrow n - m \leq (n - \rho) + (m - \rho) = m + n - 2\rho \\
& \text{and } \ m - n \leq (m - \rho) + (n - \rho) = m + n - 2\rho \\
& \Longrightarrow ¦m - n¦ \leq m + n - 2\rho \Longrightarrow 2\rho \leq m + n - ¦m - n¦ \\
(!) \qquad & \Longrightarrow \rho \leq (m + n - ¦m - n¦)/2 \Longrightarrow ¦m - n¦ \leq m + n - 2\rho \\
& \Longrightarrow m - n \leq m + n - 2\rho \ \text{ and } \ n - m \leq m + n - 2\rho \\
& \Longrightarrow 2\rho \leq 2n \ \text{ and } \ 2\rho \leq 2m \\
& \Longrightarrow m \geq \rho \ \text{ and } \ n \geq \rho .
\end{aligned}$$

Comparing the first and last lines we see that all the statements are equivalent. In particular, ρ is a lower bound for both m and n if and only if it is one of the lower bounds for $(m + n - \rceil m - n\lceil)/2$—see (!). Among the latter there is a largest, $(m + n - \rceil m - n\lceil)/2$ itself; whence the formula

$$m \wedge n = \frac{1}{2}(m + n - \rceil m - n \lceil) . \tag{6.1}$$

Since $\rho \geq m, n \iff -\rho \leq -m, -n \iff -\rho \geq (-m) \wedge (-n)$, $-((-m) \wedge (-n))$ is the least upper bound of m, n:

$$m \vee n = -\big((-m) \wedge (-n)\big) . \tag{6.2}$$

Thus $\mathfrak{M}[\mathcal{E}]$ is a vector lattice. For later use let us establish a formula for the value of $m \wedge n$ at a positive elementary integrand $\psi \in \mathcal{E}$.

Lemma 6.1 *For any* $\psi \in \mathcal{E}_+$

$$m \wedge n(\psi) = \inf \{m(\phi_1) + n(\phi_2) : \ \phi_i \in \mathcal{E}_+, \, \phi_1 + \phi_2 = \psi\} . \tag{6.3}$$

Proof. We get from Equation (6.1), using part (i) of Lemma 5.2 on p. 138 in the second line below,

$$\begin{aligned}
m \wedge n(\psi) &= \frac{1}{2}\Big(m(\psi) + n(\psi) - \rceil m - n \lceil (\psi)\Big) \\
&= \frac{1}{2}\Big(m(\psi) + n(\psi) \\
&\qquad - \sup \{(m-n)(\phi_1) - (m-n)(\phi_2) : \ \phi_i \in \mathcal{E}_+, \, \phi_1 + \phi_2 = \psi\}\Big) \\
&= \frac{1}{2}\Big(m(\psi) + n(\psi) \\
&\qquad + \inf \{(n-m)(\phi_1) + (m-n)(\phi_2) : \ \phi_i \in \mathcal{E}_+, \, \phi_1 + \phi_2 = \psi\}\Big) \\
&= \frac{1}{2} \inf\Big\{m(\phi_1) + m(\phi_2) + n(\phi_1) + n(\phi_2) \\
&\qquad + (n-m)(\phi_1) + (m-n)(\phi_2) : \ \phi_i \in \mathcal{E}_+, \, \phi_1 + \phi_2 = \psi\Big\} \\
&= \frac{1}{2} \inf\Big\{2m(\phi_2) + 2n(\phi_1) : \ \phi_i \in \mathcal{E}_+, \, \phi_1 + \phi_2 = \psi\Big\} \\
&= \inf \{m(\phi_1) + n(\phi_2) : \ \phi_i \in \mathcal{E}_+, \, \phi_1 + \phi_2 = \psi\} .
\end{aligned}$$

∎

Exercise 6.2* For two measures μ, ν of finite variation on a ring $\mathcal{A}$ of sets establish the formulae

$$\begin{aligned}
\mu \wedge \nu &= \frac{1}{2}(\mu + \nu - \rceil \mu - \nu \lceil) ; \\
\mu \wedge \nu(A) &= \inf\{\mu(A_1) + \nu(A_2) : \ A_i \in \mathcal{A}, \, A_1 + A_2 = A\} , \qquad A \in \mathcal{A} .
\end{aligned}$$

Proposition 6.3 *$\mathfrak{M}[\mathcal{E}]$ is an order complete vector lattice.*

Proof. We know that $\mathfrak{M}[\mathcal{E}]$ is a vector lattice. To finish the proof it is left to be shown that a non–void order bounded subset $\mathfrak{B} \subset \mathfrak{M}_+[\mathcal{E}]$ that is closed under taking finite suprema has a least upper bound (Lemma IV.3.7 on p. 110). Let b be any upper bound for $\mathfrak{B}$ and set

$$n(\psi) = \sup\{m(\psi) :\ m \in \mathfrak{B}\} , \qquad \psi \in \mathcal{E}_+ .$$

This supremum is finite, in fact, is less than $b(\psi)$. It is easily seen to be positive–homogeneous. It is also additive in ψ. Indeed, let $\psi_1, \psi_2 \in \mathcal{E}_+$ and $\epsilon > 0$ be given. There are $m_1, m_2 \in \mathfrak{B}$ with $|n(\psi_i) - m_i(\psi_i)| < \epsilon/3$. There is also an $m_+ \in \mathfrak{B}$ with $|n(\psi_1 + \psi_2) - m_+(\psi_1 + \psi_2)| < \epsilon/3$. With $m = m_1 \vee m_2 \vee m_+ \in \mathfrak{B}$ we have $|n(\psi_i) - m(\psi_i)| < \epsilon/3$ and $|n(\psi_1 + \psi_2) - m(\psi_1 + \psi_2)| < \epsilon/3$. Consequently

$$\begin{aligned}\Big|n(\psi_1 + \psi_2) - \Big(n(\psi_1) + n(\psi_2)\Big)\Big| &\le |n(\psi_1 + \psi_2) - m(\psi_1 + \psi_2)| \\ &\quad + \Big|m(\psi_1 + \psi_2) - \Big(m(\psi_1) + m(\psi_2)\Big)\Big| \\ &\quad + |m(\psi_1) - n(\psi_1)| + |m(\psi_2) - n(\psi_2)| \\ &\le \epsilon/3 + \epsilon/3 + \epsilon/3 = \epsilon .\end{aligned}$$

Since $\epsilon > 0$, the additivity follows. As in the proof of Lemma 5.2 (ii) we see that there is a unique extension of n to a positive elementary integral on all of $\mathcal{E}$. This is clearly a least upper bound of $\mathfrak{B}$. ∎

Some Identities. Here are a few computational rules that will come in handy from time to time. They make sense and hold on an arbitrary vector lattice $\mathfrak{M}$.

(i) For $m, n \in \mathfrak{M}$
$$m \wedge n = \frac{1}{2}(m + n - ¦m - n¦) . \tag{6.4}$$

This was shown above to be true on $\mathfrak{M}[\mathcal{E}]$. The argument did not refer to the nature of the elements of $\mathfrak{M}[\mathcal{E}]$, merely to the linear and order structure of $\mathfrak{M}[\mathcal{E}]$, so it holds for any vector lattice.

(ii) For $m, n \in \mathfrak{M}$
$$m \wedge n + m \vee n = m + n . \tag{6.5}$$

Indeed, since the shifts $z \to z+m$ and $z \to z+n$ are order isomorphisms of $\mathfrak{M}$, we have $m \wedge n + m \vee n - m - n = 0 \wedge (n-m) + (m-n) \vee 0 = 0 \wedge (n-m) - (n-m) \wedge 0 = 0$.
(iii) If we define the positive and negative parts of an element $m \in \mathfrak{M}$ by

$$m_+ = m \vee 0 \quad \text{and} \quad m_- = (-m) \vee 0 = -(m \wedge 0) ,$$

(which are both positive!) then Equation (6.5) gives

$$m = m_+ - m_- \,, \qquad m_+ \wedge m_- = 0 \,, \qquad \text{and} \qquad \lvert m \rvert = m_+ + m_- \,.$$

To see the last equality observe that $-m_- = m \wedge 0 = (m_+ - m_- - \lvert m \rvert)/2$, which implies $-m_- - m_+ = -\lvert m \rvert$. Since $m_+ \vee m_-$ majorizes both m and $-m$ and is smaller than $-m_- - m_+ = -\lvert m \rvert$, it must equal $-m_- - m_+$, and Equation (6.5) gives $m_+ \wedge m_- = 0$.
(iv) There is a ***triangle inequality*** for the variation: For $m, n \in \mathfrak{M}$,

$$\lvert m + n \rvert \leq \lvert m \rvert + \lvert n \rvert \,. \tag{6.6}$$

Indeed, $m + n = m_+ - m_- + n_+ - n_- \leq m_+ + m_- + n_+ + n_- = \lvert m \rvert + \lvert n \rvert$, and similarly $-(m + n) \leq \lvert m \rvert + \lvert n \rvert$.
(v) For $m_1, m_2, n \in \mathfrak{M}_+$,

$$(m_1 + m_2) \wedge n \leq m_1 \wedge n + m_2 \wedge n \,. \tag{6.7}$$

Indeed, $m_1 \wedge n + m_2 \wedge n = (m_1 + (m_2 \wedge n)) \wedge (n + (m_2 \wedge n)) = (m_1 + m_2) \wedge (m_1 + n) \wedge (n + m_2) \wedge (n + n) \geq (m_1 + m_2) \wedge n$.
(vi) Let $\mathfrak{B} \subset \mathfrak{M}$ have least upper bound $\bigvee \mathfrak{B}$, and let $n \in \mathfrak{M}$. Then $\mathfrak{B} \wedge n = \{m \wedge n : m \in \mathfrak{B}\}$ has least upper bound

$$\bigvee \Big(\mathfrak{B} \wedge n \Big) = \bigvee \mathfrak{B} \ \wedge n \,. \tag{6.8}$$

Indeed, the inequality $\bigvee \mathfrak{B} \ \wedge n \geq \bigvee (\mathfrak{B} \wedge n)$ is obvious. For the converse observe that

$$\begin{aligned} \bigvee \mathfrak{B} \ \wedge n + \bigvee \mathfrak{B} \ \vee n &= \bigvee \mathfrak{B} + n = \bigvee \Big(\mathfrak{B} + n \Big) \\ &= \bigvee \{ m \wedge n + m \vee n : \ m \in \mathfrak{B} \} \\ &\leq \bigvee \Big(\mathfrak{B} \wedge n \Big) + \bigvee \Big(\mathfrak{B} \vee n \Big) \\ &\leq \bigvee \Big(\mathfrak{B} \wedge n \Big) + \bigvee \mathfrak{B} \ \vee n \,, \end{aligned}$$

which results in the desired inequality $\bigvee \mathfrak{B} \ \wedge n \leq \bigvee (\mathfrak{B} \wedge n)$ upon subtraction of $\bigvee \mathfrak{B} \ \wedge n$.

Ideals and Bands. The following notions continue to make sense on an arbitrary vector lattice $\mathfrak{M}$. Two elements m, n of $\mathfrak{M}$ are called ***orthogonal*** or ***disjoint*** if $\lvert m \rvert \wedge \lvert n \rvert = 0$. In this case we write $m \perp n$. Let $\mathfrak{S} \subset \mathfrak{M}$. The set of $n \in \mathfrak{M}$ that are disjoint from every $m \in \mathfrak{S}$ will be denoted by $\mathfrak{S}^\perp$.
A collection $\mathfrak{S} \subset \mathfrak{M}$ is ***solid*** if $n \in \mathfrak{S}$ and $\lvert m \rvert \leq \lvert n \rvert$ imply $m \in \mathfrak{S}$. A solid linear subspace of $\mathfrak{M}$ is an ***ideal***. An ideal in an order complete vector lattice is a

band if it contains with every non–void order bounded set its least upper bound (and then its greatest lower bound). A band is an order complete vector lattice in its own right, but an ideal that is order complete is not necessarily a band. For instance, the bounded functions on some set S form an ideal of the vector lattice $\mathbb{R}^S$ of all finite functions, and an order complete one at that, yet they do not form a band of $\mathbb{R}^S$.

The intersection of any number of ideals (bands) evidently forms an ideal (band). We may therefore talk about the ideal (band) generated by an arbitrary non–void collection $\mathfrak{S} \subset \mathfrak{M}$. It is the intersection of all ideals (bands) containing $\mathfrak{S}$. Let us denote by $(\mathfrak{S})$ the band generated by $\mathfrak{S}$.

Theorem 6.4 (Riesz) *Let $\mathfrak{S} \subset \mathfrak{M}$. Then $\mathfrak{S}^{\perp}$ is a band. $\mathfrak{S}^{\perp\perp}$ is the band $(\mathfrak{S})$ generated by $\mathfrak{S}$. Every $m \in \mathfrak{M}$ has a unique decomposition $m = m^{\|} + m^{\perp}$ with $m^{\|} \in (\mathfrak{S})$ and $m^{\perp} \in \mathfrak{S}^{\perp}$.*

Proof. Let $m \in \mathfrak{S}^{\perp}$ and $r \in \mathbb{R}$. There exists $k \in \mathbb{N}$ with $|r| \le k$. For any $n \in \mathfrak{S}$ we have

$$\begin{aligned} \lvert rm \rvert \wedge \lvert n \rvert &= (|r| \lvert m \rvert) \wedge \lvert n \rvert \le (k \lvert m \rvert) \wedge \lvert n \rvert \\ &\le (k \lvert m \rvert) \wedge (k \lvert n \rvert) = k(\lvert m \rvert \wedge \lvert n \rvert) = 0 \,. \end{aligned}$$

Thus $\mathfrak{S}^{\perp}$ is closed under scalar multiplication. Next let $m_1, m_2 \in \mathfrak{S}^{\perp}$. Using Inequalities (6.6) and (6.7), we find

$$\lvert m_1 + m_2 \rvert \wedge \lvert n \rvert \le (\lvert m_1 \rvert + \lvert m_2 \rvert) \wedge \lvert n \rvert \le \lvert m_1 \rvert \wedge \lvert n \rvert + \lvert m_2 \rvert \wedge \lvert n \rvert = 0$$

for any $n \in \mathfrak{S}$: $\mathfrak{S}^{\perp}$ is a subspace. Its solidity is obvious, so it is an ideal. Finally, let $\mathfrak{S}$ be a non–void subset of $\mathfrak{S}^{\perp}$ that consists of positive elements and has greatest lower bound $\bigvee \mathfrak{S}$. By Equation (6.8),

$$\bigvee \mathfrak{S} \wedge \lvert n \rvert = \bigvee \{ m \wedge \lvert n \rvert : m \in \mathfrak{S} \} = 0$$

for any $n \in \mathfrak{S}$, so that $\mathfrak{S}^{\perp}$ is, indeed, a band.
For the remaining two claims observe first that $\mathfrak{S}^{\perp\perp}$ is a band containing $\mathfrak{S}$, so that $\mathfrak{S}^{\perp\perp} \supset (\mathfrak{S})$ and $\mathfrak{S}^{\perp} \cap (\mathfrak{S}) = \{0\}$. Given a positive m set

$$m^{\|} = \bigvee \{ m' \in (\mathfrak{S}) : m' \le m \} \quad \text{and} \quad m^{\perp} = m - m^{\|} \,.$$

Clearly $m^{\|} \in (\mathfrak{S})$ and $m^{\perp} \ge 0$. We have to show that $m^{\perp} \in \mathfrak{S}^{\perp}$. To do this let $n \in \mathfrak{S}$ and set

$$b = m^{\perp} \wedge \lvert n \rvert = (m - m^{\|}) \wedge \lvert n \rvert \,.$$

Then $b \in (\mathfrak{S})_+$ and $b + m^{\|} \le m$. This implies $b + m^{\|} \in (\mathfrak{S})$, $b + m^{\|} \le m^{\|}$, and $b = 0$.

If m is arbitrary we decompose its positive and negative parts into their components in $(\mathfrak{S})$ and $\mathfrak{S}^\perp$. The uniqueness of the resulting decomposition of m follows from our earlier observation that $\mathfrak{S}^\perp \cap (\mathfrak{S}) = \{0\}$. ∎

Example 6.5 Let us compute the decomposition in the particular case that $\mathfrak{S}$ consists of a single element n and that m is positive. In that case

$$m^{\|} = \bigvee_{k \in \mathbb{N}} m \wedge (k \,\vert n\vert) \,. \tag{6.9}$$

This element of $\mathfrak{M}$ evidently belongs to the band $(n) \stackrel{\text{def}}{=} (\{n\})$. We may without loss of generality assume that n is positive. It evidently suffices to show this: if $m \in (n)$ then it equals $m' \stackrel{\text{def}}{=} \bigvee_k m \wedge (kn)$. Now by (6.8) $m' + (m - m') \wedge n = m \wedge (n + m') = \bigvee_k m \wedge (n + kn \wedge m) \leq \bigvee_k m \wedge (k+1)n = m'$. This shows that $m - m' \in n^\perp$ and thus $m = m'$.

The simple situation just contemplated occurs sufficientlly often to deserve a name. We say that $m \in \mathfrak{M}$ is ***absolutely continuous with respect to*** $n \in \mathfrak{M}$, and write $m \ll n$, if m belongs to the band (n) generated by n.

Proposition 6.6 *The σ–additive elementary integrals of finite variation form a band $\mathfrak{M}^*[\mathcal{E}]$ in $\mathfrak{M}[\mathcal{E}]$. The order–continuous elementary integrals of finite variation form a band $\mathfrak{M}^\bullet[\mathcal{E}] \subset \mathfrak{M}^*[\mathcal{E}]$.*

Proof. Inspection shows that $\mathfrak{M}^*[\mathcal{E}]$ is an ideal. For instance, if $m, n \in \mathfrak{M}^*[\mathcal{E}]$ then $\vert m + n\vert$ is majorized by the sum $\vert m\vert + \vert n\vert$, which is σ–additive; it is then evidently σ–additive itself. To show that $\mathfrak{M}^*[\mathcal{E}]$ is a band, let $\mathfrak{S} \subset \mathfrak{M}^*[\mathcal{E}]$ be order bounded above with greatest lower bound $n = \bigvee \mathfrak{S}$. Without loss of generality we may assume that $\mathfrak{S}$ is increasingly directed and contained in $\mathfrak{M}^\sigma_+[\mathcal{E}]$ (Lemma IV.3.7 on p. 110). Now if $\mathcal{E}_+ \ni \phi_n \uparrow \phi \in \mathcal{E}$ then

$$\bigvee\mathfrak{S}\,(\phi) = \sup_{m \in \mathfrak{S}} m(\phi) = \sup_{m \in \mathfrak{S}, n} m(\phi_n) = \sup_n \bigvee\mathfrak{S}\,(\phi_n) \ :$$

$\bigvee\mathfrak{S} \in \mathfrak{M}^*$, so that $\mathfrak{M}^*[\mathcal{E}]$ is indeed a band. ∎

Exercise 6.7* Given a ring $\mathcal{R}$ of sets, let $\mathcal{E}$ denote the lattice ring of step functions over $\mathcal{R}$, $\mathfrak{M}[\mathcal{R}]$ the measures of finite variation on $\mathcal{R}$, and $\mathfrak{M}^*[\mathcal{R}]$ the σ–additive ones. The map $\mu \to \int d\mu$ is a linear order preserving bijection from $\mathfrak{M}[\mathcal{R}]$ onto $\mathfrak{M}[\mathcal{E}]$ that maps $\mathfrak{M}^*[\mathcal{R}]$ onto $\mathfrak{M}^*[\mathcal{E}]$. Its inverse is restriction.

Let us interpret the notions of disjointness and absolute continuity for σ–*additive* elementary integrals. To do this it is convenient to extend them all, say, in Daniell's fashion. Now the spaces $\mathcal{L}^1[m]$, generally vary with $m \in \mathfrak{M}^*[\mathcal{E}]$; since we want to compare the extensions, we better restrict them all to a common domain. The $\mathcal{E}$–Baire functions come to mind. They are measurable for every $m \in \mathfrak{M}[\mathcal{E}]$.

Most of them are too large to be integrable, so we settle for the ***$\mathcal{E}$–dominated $\mathcal{E}$–Baire functions***. These are the functions in

$$\mathcal{E}^\sigma = \left\{ h \in \mathcal{E}^\Sigma : \exists \phi \in \mathcal{E}_{00} \text{ with } |h| \le \phi \right\} .$$

They are integrable for any mean and clearly form a lattice ring. Also, $\mathcal{E}^\sigma$ contains a plethora of sets; for instance, the sets $[\phi > r] \le \phi/r$ belong to $\mathcal{E}^\sigma$ whenever $\phi \in \mathcal{E}$ and $r > 0$. We denote the restriction of the integral extension of $m \in \mathfrak{M}^*[\mathcal{E}]$ to $\mathcal{E}^\sigma$ by m^σ. It is left to the reader to show that

Proposition 6.8 *The map $m \mapsto m^\sigma$ is an order preserving bijection of $\mathfrak{M}^*[\mathcal{E}]$ onto $\mathfrak{M}^*[\mathcal{E}^\sigma]$. In particular, we have*

$$¦m^\sigma¦ = ¦m¦^\sigma ,$$

$$m \perp n \iff m^\sigma \perp n^\sigma ,$$

and

$$m \ll n \iff m^\sigma \ll n^\sigma .$$

The perpendicularity and absolute continuity expressed in the last two lines have nice characterizations in terms of the extended measures m^σ and n^σ:

Proposition 6.9 (Hahn) *Let $m, n \in \mathfrak{M}^*[\mathcal{E}]$. (i) $m \perp n$ if and only if every dominated $\mathcal{E}$–Baire set B splits into the disjoint union $B = B_m \cup B_n$ with*

$$¦m^\sigma¦(B_n) = ¦n^\sigma¦(B_m) = 0 .$$

(ii) $m \ll n$ if and only if for every dominated $\mathcal{E}$–Baire set N

$$¦n^\sigma¦(N) = 0 \implies ¦m^\sigma¦(N) = 0 .$$

Proof. (i): If every dominated $\mathcal{E}$–Baire set splits as stated, then $¦m¦ \wedge ¦n¦ = 0$ by Equation (6.3) on p. 145. To show the converse implication assume then that $m \perp n$. Given B, let $\mathcal{C}$ be a maximal collection of $\mathcal{E}$–Baire sets $C \subset B$ with $¦n¦^\sigma(C) > 0$ and $¦m¦^\sigma(C) = 0$. It is at most countable, and the sets $B_n = \bigcup \mathcal{C}$, $B_m = B \setminus B_n$ meet the description. Clearly $¦m¦^\sigma(B_n) = 0$. To see that $¦n¦^\sigma(B_m) = 0$ as well, we argue by contradiction. Suppose $¦n¦^\sigma(B_m) = \epsilon > 0$. By Equation (6.3) on p. 145, there are $\phi_k, \phi'_k \in \mathcal{E}^\sigma_+$ with sum B_m and $¦n¦^\sigma(\phi_k) + ¦m¦^\sigma(\phi'_k) < 2^{-k}$. Let $C = [\liminf_k \phi'_k > 0]$. This $\mathcal{E}$–Baire subset of B_m has $¦n¦^\sigma(C) \ge \epsilon$ and $¦m¦^\sigma(C) = 0$. It is a candidate for inclusion into $\mathcal{C}$, which was maximal.
(ii): If $m \ll n$ then by Equation (6.9) $¦m¦^\sigma = \bigvee_{k \in \mathbb{N}} ¦m¦^\sigma \wedge (k¦n¦^\sigma)$. So $¦n¦^\sigma(N) = 0$ implies $¦m¦^\sigma(N) = 0$. Conversely, if this holds then the part $¦m¦^{\sigma\perp}$ disjoint from $¦n¦^\sigma$ assigns zero mass to every dominated $\mathcal{E}$–Baire set, by part (i) above, and thus vanishes. $m^\sigma \ll n^\sigma$ follows. ∎

Supplements and Additional Exercises

Fix a vector lattice $\mathfrak{M}$.

Exercise 6.10 $n \in \mathfrak{M}$ belongs to the ideal generated by $\mathfrak{S} \subset \mathfrak{M}$ if and only if there exist $m_1, \ldots, m_k \in \mathfrak{S}$ with $¦n¦ \le ¦m_1¦ + \cdots + ¦m_k¦$.

Exercise 6.11 A linear subspace $\mathfrak{S}$ of $\mathfrak{M}$ is a band if and only if (i) $m \ll n \in \mathfrak{S}$ implies $m \in \mathfrak{S}$ and (ii) whenever $\mathcal{D}$ is a subset of $\mathfrak{S}$ consisting of mutually disjoint elements then $\sum \mathcal{D} \in \mathfrak{S}$, provided this sum exists in $\mathfrak{M}$.

A linear map $\Phi : \mathfrak{M} \to \mathfrak{N}$ to another vector lattice is a ***homomorphism of vector lattices*** if $\Phi(m \wedge n) = \Phi(m) \wedge \Phi(n) \quad \forall m, n \in \mathfrak{M}$. A homomorphism does not necessarily preserve suprema of arbitrary families — the injection of $C[0,1]$ into $\mathbb{R}^{[0,1]}$ is a counterexample. If it does it is called a ***normal homomorphism***.

Exercise 6.12 The kernel of a homomorphism is an ideal, that of a normal homomorphism a band. A homomorphism that is a bijection (an isomorphism) is normal and preserves disjointness and absolute continuity.

Exercise 6.13 Given an ideal $\mathcal{I}$ of $\mathfrak{M}$ define the ***quotient order*** on $\mathfrak{M}/\mathcal{I}$ by declaring a class $m + \mathcal{I}$ to be positive if it contains a positive element. This makes the quotient a vector lattice. If $\mathfrak{M}$ is order complete and $\mathcal{I}$ a band then the quotient is order complete as well.

Example 6.14 Let S be a set and $\mathcal{E}$ the usual collection of elementary functions on S. Let x be a point of S. The ***Dirac measure*** $\delta_x : \mathcal{E} \to \mathbb{R}$ at x is the elementary integral defined by $\delta_x(\phi) = \phi(x)$. An elementary integral m is called ***atomic*** if it belongs to the band generated by the Dirac measures; if $\mathcal{E}$ is σ–finite this is the same as saying that m is a countable linear combination of Dirac–measures: $m(\phi) = \sum_{n=1}^{\infty} r_n \phi(x_n)$ with $\sum_{n=1}^{\infty} |r_n| < \infty$.

Exercise 6.15 Show that m has finite variation and is order continuous. Compute the variation of m and show that it is order continuous.

Exercise 6.16 In the Riesz space L^p, $\dot{f} \ll \dot{g}$ if and only if every $f \in \dot{f}$ vanishes almost everywhere on every set $[g \neq 0]$, $g \in \dot{g}$.

V.7 Measures with Densitites

Let $n = \int dn$ be a σ–finite positive σ–additive elementary integral on the lattice ring $\mathcal{E}$ of bounded functions, and $\|\ \|_n^*$ its Daniell mean. A function g is called ***locally n-integrable*** if $\phi \cdot g$ is n–integrable for every $\phi \in \mathcal{E}$. If n is Lebesgue measure on the line then g is locally integrable if and only if it is integrable on every bounded interval. Given such g, consider the map

$$\phi \mapsto m(\phi) \stackrel{\text{def}}{=} \int \phi \cdot g \, dn , \qquad \phi \in \mathcal{E} .$$

This is plainly (why?) a σ–additive elementary integral on $\mathcal{E}$. It is called ***the measure with base n and density g***. We write $m = gn$. m has finite variation, since for all $\phi \in \mathcal{E}$

$$|m(\phi)| \leq \int |\phi| \cdot |g|\, dn \ ,$$

which says that $\int |g| dn$ majorizes both m and $-m = \int (-g) dn$. In other words,

$$\vert m \vert = \vert gn \vert \leq |g| n \ . \tag{7.1}$$

There is actually equality, but we defer showing this. Note now though that

$$\int \phi \cdot |g| dn = \sup_k \int \phi \cdot (|g| \wedge k) dn$$

for $\phi \in \mathcal{E}_+$. Since the $(|g| \wedge k)n$ are majorized by multiples of n, they belong to the ideal generated by n. $|g|n$, being the greatest lower bound of the family $(|g| \wedge k)n$, belongs to the band generated by n, and then so does m. In other words, m is absolutely continuous with respect to n.

Exercise 7.1 This can be seen also by applying Proposition 6.9 on p. 150 (ii), once it is shown that $m^\sigma(h) = \int fg\, dn$ for $h \in \mathcal{E}^\sigma$.

There are natural conjectures: f is $\| \ \|_m^*$–integrable (–negligible, –measurable) if and only if the product fg is $\| \ \|_n^*$–integrable (–negligible, –measurable), etc. They are true. To prove them it is convenient to consider the mean $f \mapsto \|f\|_m^\flat = \|fg\|_n^*$. Namely, it is plain that f is $\| \ \|_m^\flat$–integrable if and only if fg is $\| \ \|_n^*$–integrable. So if we can show that $\| \ \|_m^\flat$ coincides with the Daniell mean $\| \ \|_m^*$ for $\vert m \vert$ then we have access to a trivial proof of the conjecture. Note that $\|f\|_m^* \leq \|f\|_m^\flat$ for $\mathcal{E}$–Baire functions inasmuch as this inequality is satisfied on $\mathcal{E}_+$ (Proposition III.5.7 on p. 88).

Lemma 7.2 $\|f\|_m^* \leq \|f\|_m^\flat$.

Proof. There is an increasing sequence (ϕ_k) of positive elementary integrands with pointwise supremum 1 (Exercise II.7.2 on p. 62). Let γ_k be an $\mathcal{E}$–Baire upper envelope of the integrable function $|g| \cdot [\phi_k > 1/k]$ (Proposition III.6.2 on p. 93), and set $\gamma = \liminf_k \gamma_k$. This is an $\mathcal{E}$–Baire function that exceeds $|g|$ and agrees $\|f\|_n^*$–almost everywhere with $|g|$. Clearly

$$\|f\|_m^\flat = \|f\gamma\|_n^* \ .$$

Next observe that for any f and $k \in \mathbb{N}$ there is an $\mathcal{E}$–Baire function $h_k \geq (f \wedge k) \cdot \phi_k[\gamma = 0]$ with $\|h_k\|_m^* < \infty$. We may choose h_k so it vanishes on $[\gamma > 0]$.

We have $\quad \big\|f \wedge k \cdot \phi_k \cdot [\gamma = 0]\big\|_m^* \leq \|h_k\|_m^* \leq \|h_k\|_m^\flat = \|h_k \cdot \gamma\|_n^* = 0 \ ,$

and consequently $\quad \big\|f \cdot [\gamma = 0]\big\|_m^* = \sup_k \big\|f \wedge k \cdot \phi_k \cdot [\gamma = 0]\big\|_m^* = 0 \ .$

Clearly $\quad \big\|f \cdot [\gamma = 0]\big\|_m^\flat = 0$

as well. In comparing $\|f\|_m^*$ and $\|f\|_m^\flat$ we may therefore assume that f vanishes on $[\gamma = 0]$. Suppose then that $\|f\|_m^\flat < r$ for such an f. There is an $\mathcal{E}$–Baire function $h \geq |f|\gamma$ with $\|h\|_n^* < r$. Then $f \leq h[\gamma > 0]/\gamma$ and consequently

$$\|f\|_m^* \leq \big\|h[\gamma > 0]/\gamma\big\|_m^* \leq \big\|h[\gamma > 0]/\gamma\big\|_m^\flat \leq \|h\|_n^* < r \,.$$

Theorem 7.3 *The variation of $m = gn$ is $|g|n$, and $\|f\|_m^\flat = \|f\|_m^*$. The function f is $\|\ \|_m^*$–integrable (–measurable, –negligible) if and only if the product function fg is $\|\ \|_n^*$–integrable (–measurable, –negligible).*

Proof. Suppose f is $\|\ \|^\flat$–integrable. There are $\phi_k \in \mathcal{E}$ with $\|(f - \phi_k)g\|_n^* \to 0$. As $\phi_k g$ is $\|\ \|_n^*$–integrable, fg is $\|\ \|_n^*$–integrable.
Let us show next that $\psi[g \neq 0]/g$ is $\|\ \|^\flat$–integrable, for $\psi \in \mathcal{E}$. Indeed, this function is clearly $\|\ \|_n^*$–measurable, and since evidently $\|\ \|_m^\flat \ll \|\ \|_n^*$, it is also $\|\ \|_m^\flat$–measurable (Corollary III.3.7 on p. 77). It has finite $\|\ \|_m^\flat$–mean, so is $\|\ \|_m^\flat$–integrable (Theorem III.3.1 on p. 75).
Assume then that fg is $\|\ \|_n^*$–integrable. Then there are $\psi_k \in \mathcal{E}$ with

$$\|fg - \psi_k\|_n^* = \|(f - \psi_k[g \neq 0]/g) \cdot g\|_n^* = \|f - \psi_k[g \neq 0]/g\|_m^\flat \to 0 \,.$$

By the previous argument, f is $\|\ \|_m^\flat$–integrable. We have shown that f is $\|\ \|_m^\flat$–integrable if and only if fg is $\|\ \|_n^*$–integrable. That f is $\|\ \|_m^\flat$–negligible if and only if fg is $\|\ \|_n^*$–negligible is obvious.
Let us go after measurability. If fg is $\|\ \|_n^*$–measurable then so is $f[g \neq 0] = fg[g \neq 0]/g$. This function is therefore $\|\ \|_m^\flat$–measurable. Since $f[g = 0]$ is $\|\ \|_m^\flat$–negligible it is $\|\ \|_m^\flat$–measurable, and then so is the sum f. Conversely, assume f is $\|\ \|_m^\flat$–measurable. Since $\|\ \|_m^\flat$ and $f' \mapsto \|f'[g \neq 0]\|_n^*$ have the same negligible sets, they have the same measurable functions (Corollary III.3.7 on p. 77). We leave it to the reader to show that this implies that f is $\|\ \|_n^*$–measurable on the set $[g \neq 0]$. Clearly fg is then $\|\ \|_n^*$–measurable there as well. It is also $\|\ \|_n^*$–measurable on $[g = 0]$, since it vanishes there: fg is $\|\ \|_n^*$–measurable.
After these preparations let us attack the first claim of the theorem. By the very definition of the variation,

$$\int \phi \, d\mu = \int \phi g \, dn \leq \int \phi \, d\,\rvert m\rvert \tag{7.2}$$

for $\phi \in \mathcal{E}_+$. The function $\phi[g > 0]$ is $\|\ \|_m^\flat$–integrable. There are $\phi_k \in \mathcal{E}_+$ that converge in $\|\ \|_m^\flat$–mean, and by the lemma in $\|\ \|_m^*$–mean to $\phi[g > 0]$. We apply Inequality (7.2) and take the limit. By continuity

$$\int \phi[g > 0]g \, dn \leq \int \phi[g > 0] \, d\,\rvert m\rvert \,.$$

The same argument applies to $-g$ and yields

$$-\int \phi[g<0]g\,dn \le \int \phi[g<0]\,d¦m¦\;.$$

Upon addition
$$\int \phi|g|\,dn \le \int \phi\,d¦m¦$$

follows. The reverse inequality being obvious, equality obtains.
This implies that $\|\ \|_m^\flat = \|\ \|_m^*$ on $\mathcal{E}_+$ and, since $\|\ \|_m^*$ is maximal, that $\|\ \|_m^\flat \le \|\ \|_m^*$. The lemma produces equality. The first part of the proof yields the remaining claims. ∎

Exercise 7.4 Show that $g(x) = e^{x^2}$ is a locally Lebesgue integrable function. Find a Lebesgue integrable set A such that Ag is not Lebesgue integrable.

V.8 The Radon–Nikodym Theorem

In the last section we have produced a large number of elementary integrals on $\mathcal{E}$ simply by multiplying a given $n \in \mathfrak{M}^*[\mathcal{E}]$ with a locally n–integrable density g. The result was a new elementary integral $g \cdot n$ absolutely continuous with respect to n. If n is σ–finite, every $m \ll n$ can be obtained in this way:

Theorem 8.1 (Radon–Nikodym) *Suppose n is a σ–finite σ–additive positive elementary integral on the lattice ring $\mathcal{E}$, and m a σ–additive elementary integral of finite variation. The following are equivalent:*
(i) m is absolutely continuous with respect to n;
(ii) for every $\mathcal{E}$–Baire set N, $\int N\,dn = 0$ implies $\int N\,dm = 0$;
(iii) There exists a locally $\|\ \|_n^*$–integrable density g such that $m = gn$.

In this case g is called a ***Radon–Nikodym derivative*** *of m with respect to n and one writes $g = dm/dn$.*

Proof. There is a sequence (B_k) of dominated $\mathcal{E}$–Baire sets whose union is the ambient set.
(i) implies (ii): Assume $\int N\,dn = 0$. Then $\int B_k N\,dm = 0$ for all $k \in \mathbb{N}$ by Proposition 6.9 on p. 150 (ii), and therefore $\int N\,dm = 0$. The implication (i) $\Longrightarrow$ (ii) follows similarly. That (iii) implies (i) we have seen in Theorem 7.3 on p. 153.
Assume then that $m \ll n$. Since $m_\pm \le ¦m¦$, we have $m_\pm \ll n$, and it suffices to produce a density for the positive and negative parts of m separately. In other words, we may assume that m is positive.

For every $k \in \mathbb{N}$ let $m_k = m \wedge kn$ be the infimum of the elementary integrals m and kn. From Equation (6.3) on p. 145 we get the following formula for m_k: for any $\psi \in \mathcal{E}_+$,

$$m_k(\psi) = \inf \{ m(\phi_1) + kn(\phi_2) : \ \phi_i \in \mathcal{E}_+, \, \phi_1 + \phi_2 = \psi \} \le k \cdot n(\psi) .$$

(The inequality comes from the choice $\phi_1 = 0$.) This says that $m_k \le kn$ and implies

$$|m_k(\phi)| \le k \cdot n(\phi) \le k \cdot \|\phi\|_n^* .$$

Thus m_k has a unique extension by continuity to a linear map $\int dm_k : \mathcal{L}^1[n] \to \mathbb{R}$. We know from Theorem IV.5.5 on p. 121 what the continuous linear functionals on $\mathcal{L}^1[n]$ are: they are integration against a function in $\mathcal{L}^\infty[n]$. There is therefore a bounded measurable function g_k such that

$$m_k(f) = \int f \cdot g_k \, dn , \qquad \forall \, f \in \mathcal{E}^\sigma .$$

Since $m_k \ge 0$ we may choose $g_k \ge 0$. Since $m_{k+1} \ge m_k$ we may choose (g_k) increasing. The desired density g is the pointwise supremum of the g_k. This is almost everywhere finite, since $\sup_k \int \psi g_k dn = \int \psi \, dm < \infty$ for all $\psi \in \mathcal{E}_+$. This formula also shows that $m = gn$. ∎

V.9 An Application: Conditional Expectation

The Image of a Measure

Let $(S, \mathcal{E}, \mu)$ be a σ–additive elementary integral, $(S', \mathcal{E}')$ a space equipped with a ring of bounded functions, and $\Phi : S \to S'$ a map satisfying the following assumption: $\phi' \circ \Phi \in \mathcal{L}^1[\mathcal{E}, \| \ \|^\mu]$ for every $\phi' \in \mathcal{E}'$. Then $\nu(\phi') \stackrel{\text{def}}{=} \int \phi' \circ \Phi \, d\mu$ exists for all $\phi' \in \mathcal{E}'$ and defines a σ–additive elementary integral $\nu = \int d\nu$ on $\mathcal{E}'$. If μ is positive then clearly so is ν. The measure ν is called the ***image of μ under Φ*** and is denoted by $\Phi[\mu]$. In other words,

$$\Phi[\mu](\phi') \stackrel{\text{def}}{=} \int \phi' \circ \Phi \, d\mu , \qquad \phi' \in \mathcal{E}' .$$

What do the integration theories of μ and $\Phi[\mu]$ have to do with each other?

Theorem 9.1 *If $f' : S' \to \overline{\mathbb{R}}$ is ν–integrable then $f' \circ \Phi$ is μ–integrable and*

$$\int f' \, d\nu = \int f' \circ \Phi \, d\mu .$$

If f' is ν–measurable then $f' \circ \Phi$ is μ–measurable.

Proof. For $g' : S' \to \overline{\mathbb{R}}$ define $\|g'\|^\flat = \|g' \circ \Phi\|^\mu$. This is a mean that agrees with $\| \ \|^\nu$ on $\mathcal{E}'$. By the maximality of Daniell's mean (Proposition III.6.2) $\| \ \|^\flat = \| \ \|^\nu$ on $\mathcal{L}^1[\nu]$. So if f' is the $\| \ \|^\nu$–mean limit of $\phi'_n \in \mathcal{E}'$ then $f' \circ \Phi$ is the $\| \ \|^\mu$–mean limt of $\phi_n \circ \Phi \in \mathcal{L}^1[\mu]$. The second claim is left to the reader. ∎

Exercise 9.2 If the set $A' \subset S'$ is ν–integrable then $\Phi^{-1}(A') \subset S$ is μ–integrable, and $\nu(A') = \mu(\Phi^{-1}(A'))$. For this reason ν is denoted by $\mu \circ \Phi^{-1}$ or even $\Phi^{-1}\mu$ by some authors.

Conditional Expectation

Assume now that μ is positive, and let $g \in \mathcal{L}^1[\mu]$. Then $g{\cdot}\mu$ is a σ–additive measure absolutely continuous with respect to μ. Consider the measure $\Phi[g{\cdot}\mu]$ on $\mathcal{E}'$. If g is positive and bounded then clearly $\Phi[g{\cdot}\mu] \ll \nu$. If merely $g > 0$ then $\Phi[g{\cdot}\mu] = \sup_n \Phi[(g \wedge n){\cdot}\mu]$ still belongs to the band (ν) generated by ν, and therefore $\Phi[g{\cdot}\mu] \ll \nu$ (Theorem 8.1). In general

$$\Phi[g{\cdot}\mu] = \Phi[g_+{\cdot}\mu] - \Phi[g_-{\cdot}\mu] \in (\nu) .$$

If $\mathcal{E}'$ is σ–finite then there exists a Radon–Nikodym derivative of $\Phi[g{\cdot}\mu]$ with respect to $\nu = \Phi[\mu]$ (*ibidem*), which is denoted $\mathbb{E}^\mu[g|\Phi]$. It is thus defined as a ν–integrable function satisfying

$$\int f' \cdot g \, d\mu = \int f' \cdot \mathbb{E}^\mu[g|\Phi] \, d\nu = \int \big(f' \cdot \mathbb{E}^\mu[g|\Phi]\big) \circ \Phi \, d\mu \qquad \forall f' \in \mathcal{L}^1_b[\nu] .$$

It is called the ***conditional expectation of g under Φ (and μ).***

In probability theory the situation often—but not exclusively—is this: the spaces S and S' coincide and $\Phi : S \to S'$ is the identity; $\mathcal{E}$ is the lattice ring of step functions over a σ–algebra $\mathcal{F}$, and $\mathcal{E}'$ the lattice ring of step functions over a sub–σ–algebra $\mathcal{F}' \subset \mathcal{F}$. $\nu = \Phi[\mu]$ is then nothing but the restriction of μ to $\mathcal{F}'$. In this case one writes $\mathbb{E}^\mu[g|\mathcal{F}']$ for $\mathbb{E}^\mu[g|\Phi]$, or simply $\mathbb{E}[g|\mathcal{F}']$ if it is clear which probability μ is meant, and talks about the conditional expectation of g under $\mathcal{F}'$. It can be chosen measurable on $\mathcal{F}'$ and is defined by

$$\int f' \cdot g \, d\mu = \int f' \cdot \mathbb{E}^\mu[g|\mathcal{F}'] \, d\mu$$

for all bounded f' measurable on $\mathcal{F}'$.

Exercise 9.3 Any two conditional expectations of g under Φ or $\mathcal{F}'$ differ ν–negligibly.

Exercise 9.4 (Jensen's Inequality) Suppose $1 \in \mathcal{E}$ and $\mu \geq 0$ has total mass $\mu(1) = 1$. (i) The map $f \mapsto \mathbb{E}^\mu[f|\Phi]$ is linear and positive, maps 1 to 1, and is contractive from

$L^p(\mu)$ to $L^p(\nu)$ when $1 \le p \le \infty$. (ii) Let $\Gamma : \mathbb{R} \to \mathbb{R}$ be a convex function such that $\Gamma \circ f$ is ν–integrable. Then ν–almost surely

$$\Gamma(\mathbb{E}^{\mu}[f|\Phi]) \le \mathbb{E}^{\mu}[\Gamma(f)|\Phi] . \tag{9.1}$$

V.10 Differentiation

From the Calculus we expect something like this: if f is Lebesgue integrable then its indefinite integral $t \mapsto \int_0^t f(s)\,ds$ is a differentiable function $F(t)$ of t whose derivative is f; if F is differentiable then its derivative $F' = f$ is integrable and its indefinite integral equals F up to a constant. The Fundamental Theorem of Calculus says that these statements are true if f or the derivative of F are *a priori* known to be continuous. In this section we investigate what can be said about more general functions f and F. The main result is this:

Theorem 10.1 *A function F of finite variation on the line is λ–almost everywhere differentiable.*

It is also true that the derivative $F' = f$ is — at least locally — integrable. It is not true in general that F is an indefinite integral of its derivative f. Below we give a proof of Theorem 10.1 that explains the deficiency between $\int^x f$ and $F(x)$ and at the same time introduces the reader to some arguments that recur throughout Analysis and Probability.

By Exercise 5.8 on p. 142, F is the distribution function of some measure μ_F of finite variation. According to Theorem 6.4 μ_F has a unique decomposition

$$\mu_F = \mu^{\|} + \mu^{\perp}$$

into a measure $\mu^{\|}$ on $\mathcal{A}[\mathbb{R}]$ absolutely continuous with respect to Lebesgue measure λ and a measure $\mu^{\perp}$ disjoint from λ. By the Radon–Nikodým Theorem 8.1, the former has a density g with respect to λ and is of the form $g \cdot \lambda$. In summary, F is of the form [1]

$$F(t) = \int_0^t g(s)\,ds + F^{\perp}(t) ,$$

where g is locally Lebesgue integrable and $F^{\perp}$ is the distribution function of a measure $\mu^{\perp}$ disjoint from Lebesgue measure. The proof of Theorem 10.1 will consist in showing two subsidiary results:
(i) the derivative of $\int_0^t g(s)\,ds$ equals $g(t)$ λ–almost everywhere, and
(ii) the derivative of $F^{\perp}$ vanishes λ–almost everywhere. This explains why F is

[1] As usual, $\int_0^t g(s)\,ds = -\int_t^0 g(s)\,ds$ for $t < 0$.

not, in general, an indefinite integral of its derivative; μ_F could be disjoint from Lebesgue measure.

Let us simplify the first statement

$$\frac{d}{dt}\int_0^t g(s)\,ds = g(t)\ \lambda\text{–a.e.} \tag{10.1}$$

if g is *locally* integrable, to the case that g is integrable: Let $g_n = g \cdot (-n, n)$, $n \in \mathbb{N}$. g_n is integrable, and $\int_0^t g_n(s)\,ds = \int_0^t g(s)\,ds$ for $t \in (-n, n)$. If we can establish that $\frac{d}{dt}\int_0^t g_n(s)\,ds = g_n(t)$ a.e. then (10.1) is clearly true for our locally integrable g. That is to say, point (i) is reduced to the following

Proposition 10.2 *Assume f is Lebesgue integrable, and set $F(t) = \int_{-\infty}^t f(s)\,ds$. Then F is uniformly continuous. Moreover, for λ–almost every $t \in \mathbb{R}$ the derivative $F'(t)$ exists and equals $f(t)$.*

Point (ii) will be taken up in Proposition 10.4 below.

Proof of Proposition 10.2. Replicas of the following arguments are basic in martingale theory, ergodic theory, and many other areas of Analysis.
For $\phi \in C_{00}[\mathbb{R}]$, $t \mapsto F_\phi(t) = \int_0^t \phi(s)\,ds$ is uniformly continuous. Given $\epsilon > 0$, we find ϕ such that $\|f - \phi\|_1 < \epsilon$ and observe that $|F(t) - F_\phi(t)| < \epsilon$ for all t. Thus F, being the uniform limit of uniformly continuous functions, is uniformly continuous itself.

Its differentiability almost everywhere is much more complicated to prove. For every $h \neq 0$ let us define the operator D_h that takes integrable functions f to continuous functions $D_h f$ by

$$D_h f(t) = \frac{1}{h}\int_t^{t+h} f(s)\,ds\ . \qquad -\infty < h < \infty$$

We want to show that $D_h f \xrightarrow[|h|\to 0]{} f$ a.e. To this end we provide an estimate of the ***Hardy–Littlewood Maximal Operator*** D^*, which is defined as the supremum instead of the limit:

$$(D^* f)(t) \stackrel{\text{def}}{=} \sup_{|h|>0} |D_h f(t)|\ .$$

Lemma 10.3 (The Hardy–Littlewood Maximal Lemma) *For any Lebesgue integrable function g and $\alpha > 0$*

$$\lambda^*([D^* g > \alpha]) \le \frac{1}{\alpha}\int_{[D^* g > \alpha]} |g(s)|\,ds\ .$$

We defer the proof of this inequality and show first how it is used to prove Proposition 10.2. Let ϕ be any continuous function of compact support. Then

$$\begin{aligned}|D_h f - f| &\le |D_h(f-\phi)| + |D_h\phi - \phi| + |\phi - f| \\ &\le D_h\,|f-\phi| + |D_h\phi - \phi| + |\phi - f|\,,\end{aligned}$$

and consequently for any $\delta > 0$

$$\sup_{0<|h|<\delta} |D_h f - f| \le \sup_{0<|h|<\delta} D_h(|f-\phi|) + \sup_{0<|h|<\delta} |D_h\phi - \phi| + \sup_{0<|h|<\delta} |\phi - f|\,.$$

Let $\alpha > 0$. Then

$$\begin{aligned}[\sup_{0<|h|<\delta} |D_h f - f| > \alpha] \subset \Big(&[\sup_{0<|h|<\delta} D_h(|f-\phi|) > \alpha/3] \\ &\cup [\sup_{0<|h|<\delta} |D_h\phi - \phi| > \alpha/3] \\ &\cup [\sup_{0<|h|<\delta} |\phi - f| > \alpha/3] \Big)\end{aligned}$$

and

$$\begin{aligned}\lambda^*\Big([\sup_{0<|h|<\delta} |D_h f - f| > \alpha]\Big) \le &\lambda^*\Big([\sup_{0<|h|<\delta} D_h(|f-\phi|) > \alpha/3]\Big) \\ &+ \lambda^*\Big([\sup_{0<|h|<\delta} |D_h\phi - \phi| > \alpha/3]\Big) \\ &+ \lambda^*\Big([|\phi - f| > \alpha/3]\Big)\,.\end{aligned} \tag{10.2}$$

Given an $\epsilon > 0$ we choose now, as we may, $\phi \in C_{00}[\mathbb{R}]$ such that $\|f-\phi\|_1 < \alpha\epsilon$. By the lemma the first summand of (10.2) is less than

$$\lambda^*([D^*(|f-\phi|) > \alpha/3]) \le \frac{1}{(\alpha/3)} \int_{[D^* f > \alpha/3]} |f-\phi| \le \frac{\alpha\epsilon}{(\alpha/3)} = 3\epsilon\,.$$

We choose now $\delta > 0$ so that $|s-t| < \delta$ implies $|\phi(s) - \phi(t)| < \alpha/3$. This is possible since ϕ is uniformly continuous. With this choice, the second summand in (10.2) actually is zero. Namely, for $|h| < \delta$

$$|D_h\phi(t) - \phi(t)| \le \frac{1}{h} \int_t^{t+h} |\phi(s) - \phi(t)|\, ds < \alpha/3 \quad \forall\, t\,,$$

so that the set $[\sup_{0<|h|<\delta} |D_h\phi - \phi| > \alpha/3]$ is void. The last summand of (10.2) is by Chebyshev's inequality (Exercise II.3.7) smaller than $\frac{3}{\alpha} \cdot \alpha\epsilon = 3\epsilon$. In summary, for this choice of δ

$$\lambda^*([\sup_{0<|h|<\delta} |D_h f - f| > \alpha]) < 6\epsilon\,.$$

A fortiori

$$\lambda^*([\limsup_{|h|\to 0} |D_h f - f| > \alpha]) < 6\epsilon\,.$$

This is true for arbitrary $\alpha, \epsilon > 0$. We conclude that the set

$$N = [\limsup_{|h|\to 0} |D_h f - f| > 0]$$

is Lebesgue negligible. For $t \notin N$ we have $D_h f(t) \xrightarrow[|h|\to 0]{} f(t)$, which is the claim of Proposition 10.2. ∎

Proof of Lemma 10.3. The set $[D^*g > \alpha]$ is clearly open, so it can be written as a countable union of disjoint open intervals (a_k, b_k). The desired equality follows upon addition from

$$b_k - a_k = \frac{1}{\alpha} \int_{a_k}^{b_k} g(s)\, ds \,, \tag{10.3}$$

which is therefore all there is to prove. Let then $t \in (a_k, b_k)$.
Scenario I: there is a number $u > t$ with

$$\int_t^u g(s)\, ds > \alpha(u - t) \,. \tag{10.4}$$

(Ia) Then numbers u satisfying (10.4) can be found in the interval (t, b_k). For if the opposite inequality were satisfied at all points of (t, b_k), then it would be satisfied at the point b_k as well, by continuity. Now since $b_k \notin [D^*g > \alpha]$ we would have $\int_{b_k}^u g(s)\, ds \le \alpha(u - b_k)$ for all $u > b_k$ and $\int_t^u g(s) ds = \int_t^{b_k} + \int_{b_k}^u \le \alpha(b_k - t + u - b_k)$ for all $u > t$. This is impossible in the present scenario.
(Ib) The supremum v of all $u \in (t, b_k)$ satisfying (10.4) is b_k. For if $v < b_k$ then $\int_t^v g(s)\, ds = \alpha(v - t)$. Redoing *(Ia)* with v replacing t would provide a $u \in (v, b_k)$ with $\int_v^u g(s)\, ds > \alpha(u - v)$, and adding the last two (in–)equalities would result in a $u > v$ satisfying (10.4), in contradiction to the definition of v. This argument shows also that b_k cannot equal ∞.
There are therefore arbitrarily close to b_k numbers $u < b_k$ that fall under
Scenario II: there is a $t < u$ satisfying (10.4). Arguments *(Ia)* and *(Ib)* suitably altered show that such t can be found arbitrarily close and to the right of a_k. We take the limit in (10.4) as $t \downarrow a_k$ and $u \uparrow b_k$ and arrive at $\int_{a_k}^{b_k} g(s)\, ds \ge \alpha(b_k - a_k)$. The inequality cannot be strict, else a_k, b_k would belong to $[D^*g > \alpha]$: Equation (10.3) holds.
Finally, if the t we started out with does not fall under scenario I, it must meet scenario II, and we repeat the argument with the order of $\mathbb{R}$ reversed: Equation (10.3) is proved in both cases. Note that this finishes the proof of Proposition 10.2. To complete the proof of Theorem 10.1 it suffices to show the following ∎

Proposition 10.4 *Let F be a function of finite variation on the line and assume that its associated measure μ_F is disjoint from Lebesgue measure λ. Then F is differentiable λ–almost everywhere and its derivative is almost everywhere zero.*

To prove this a subsidiary notion and result are needed. Let E be a subset of the line. A collection $\mathcal{V}$ of closed intervals is a ***Vitali cover*** of E if every point of E is contained in intervals of $\mathcal{V}$ of arbitrarily small but strictly positive length.

Lemma 10.5 (Vitali's Covering Theorem) *Suppose E be a subset of $\mathbb{R}$ of finite Lebesgue outer measure and $\mathcal{V}$ is a Vitali cover of E. Then $\mathcal{V}$ contains a countable disjoint family $\{I_1, I_2, \ldots\}$ that covers E λ–almost everywhere:*

$$\lambda^*(E \setminus \bigcup_{n=1}^{\infty} I_n) = 0 .$$

Proof. Let U_0 be an open set of finite outer measure containing E. It is good enough to produce the I_n from the Vitali cover $\{I \in \mathcal{V} : I \subset V\}$. In other words, we may assume that the sets of $\mathcal{V}$ all are contained in U_0.
The construction of the I_n is by induction. For I_1 pick any element of $\mathcal{V}$. Suppose $I_1, \ldots, I_n \in \mathcal{V}$ have been chosen and are mutually disjoint. If $E \subset \bigcup_{k=1}^{n} I_k$ the construction is complete. Otherwise set

$$U_n = U_0 \setminus \bigcup_{k=1}^{n} I_k .$$

This set is clearly open. Set

$$\delta_n = \sup\{\lambda(I) : I \in \mathcal{V}, I \subset U_n\} .$$

This number is strictly positive, since there are a point $x \in E$ not in $\bigcup_{k=1}^{n} I_k$ and an interval of $\mathcal{V}$ not intersecting this set. Pick then $I_{n+1} \in \mathcal{V}$ such that $I_{n+1} \cap \bigcup_{k=1}^{n} I_k = \emptyset$ and $\lambda(I_{n+1}) > \delta_n/2$. The construction of $I_1, I_2, \ldots$ is complete. It is to be shown that $\lambda^*(E \setminus \bigcup_{n=1}^{\infty} I_n) = 0$. For each n let J_n denote the interval that has the same midpoint as I_n but five times its length. Then

$$\lambda^*(\bigcup_{n=1}^{\infty} J_n) \leq \sum_{n=1}^{\infty} \lambda(J_n) \leq 5 \sum_{n=1}^{\infty} \lambda(I_n) \leq 5\lambda(U_0) < \infty .$$

Thus

$$\lim_{M \to \infty} (\bigcup_{n=M}^{\infty} J_n) = 0 .$$

It suffices therefore to show that

$$\left(E \setminus \bigcup_{n=1}^{\infty} I_n\right) \subset \bigcup_{n=M}^{\infty} J_n \qquad \forall M \in \mathbb{N} . \tag{10.5}$$

Fix then $M \in \mathbb{N}$. Now if $x \in E \setminus \bigcup_{n=1}^{\infty} I_n$, then $x \in U_M$, and there exists an $I \in \mathcal{V}$ with $x \in I \subset U_M$. Since $\delta_n < 2\lambda(I_{n+1}) \xrightarrow[n\to\infty]{} 0$, there exists an $n \in \mathbb{N}$ such that $\delta_n < \lambda(I)$ and consequently $I \not\subset U_n$. Let N be the smallest integer such that $I \not\subset U_n$. Clearly $M < N$. Since I does not intersect $\bigcup_{k=1}^{N-1} I_k$ but $\bigcup_{k=1}^{N} I_k$, its intersection with I_N is non–void. As $\lambda(I) \leq \delta_{N-1} < 2\lambda(I_N)$,

$$I \subset J_N \subset \bigcup_{n=M}^{\infty} J_n . \qquad \blacksquare$$

Exercise 10.6* Show that the conclusion holds even if E has infinite outer measure.

Proof of Proposition 10.4: Decomposing the measure μ_F into its positive and negative parts, and with that writing F as the difference of two increasing functions whose associated measures are disjoint from Lebesgue measure λ, we reduce the situation to the case that F is increasing. Write $\mu = \mu_F$. It is to be shown that

$$\limsup_{|h|\to 0} \frac{1}{h}(F(t+h) - F(t)) = 0 \tag{10.6}$$

λ–almost everywhere. It clearly suffices to show that (10.6) holds at λ–almost all $t \in (-M, M]$, for every $M \in \mathbb{N}$. Fix then an $M \in \mathbb{N}$. Let $0 < \alpha \leq 1$ and consider the set

$$E_\alpha^M = \left\{ t \in (-M, M] : \limsup_{|h|\to 0} \frac{F(t+h) - F(t)}{h} > \alpha \right\} .$$

We shall show that

$$\lambda^*(E_\alpha^M) < \epsilon , \tag{10.7}$$

for any $\epsilon > 0$. This clearly implies Equation (10.6) for almost all $t \in (-M, M]$, and we will be done.

Given $\epsilon > 0$, we use Exercise 6.2 on p. 145 to write the elementary interval $(-M, M]$ as the disjoint union of two elementary sets A^λ and A^μ in $\mathcal{A}[\mathbb{R}]$ with

$$\lambda(A^\mu) + \mu(A^\lambda) < \alpha\epsilon/2 .$$

For the part of E_α^M in A^μ we get the simple estimate

$$\lambda^*\left(E_\alpha^M \cap A^\mu\right) < \epsilon/2 . \tag{10.8}$$

Since A^λ differs from its interior $\mathring{A}^\lambda$ in at most finitely many points, we have $\lambda^*\left(E_\alpha^M \cap A^\mu\right) = \lambda^*\left(E_\alpha^M \cap \mathring{A}^\mu\right)$. For every point t in $E_\alpha^M \cap \mathring{A}^\mu$ there are arbitrarily small intervals of the form $I = (t, t+h]$ or $I = (t-h, t]$ with $\mu(I) > \alpha\lambda(I) =$

$\alpha\lambda(\overline{I})$. The closures $\overline{I}$ of these intervals form a Vitali cover of $E_\alpha^M \cap \mathring{A}^\mu$. Using Vitali's covering Lemma 10.5, we get a sequence (I_n) of such intervals with disjoint closures and

$$\begin{aligned}\lambda^*\left(E_\alpha^M \cap A^\mu\right) &= \lambda^*\left(E_\alpha^M \cap \mathring{A}^\mu\right)\\ &\le \sum_n \lambda(\overline{I}_n) = \sum_n \lambda(I_n)\\ &< \frac{1}{\alpha}\sum_n \mu(I_n) \le \frac{1}{\alpha}\cdot \mu(A^\lambda) < \epsilon/2\,.\end{aligned}$$

We add this inequality to Inequality (10.8) and obtain Inequality (10.7). ∎

Supplements and Additional Exercises

We prove here the famous $\mathcal{L}^p$–inequalities for the Hardy–Littlewood Maximal Operator D^*. Throughout, f is a locally Lebesgue integrable function.

Exercise 10.7* Lemma 10.3 persists if g is merely locally integrable.

Lemma 10.8 *For any $\alpha > 0$ and $\beta \in (0,1)$*

$$\lambda^*([D^*f > \alpha]) \le \frac{1}{\alpha(1-\beta)}\cdot\int f[f \ge \alpha\beta]\,d\lambda\,. \tag{10.9}$$

Proof. The inequalities

$$\begin{aligned}D^*f &\le D^*\big(g\cdot[g>\alpha\beta]\big) + D^*\big(g\cdot[g\le\alpha\beta]\big)\\ &\le D^*\big(g\cdot[g>\alpha\beta]\big) + \alpha\beta\end{aligned}$$

show that $\quad \big[D^*g > \alpha\big] \subset \big[D^*(g\cdot[g>\alpha\beta]) > \alpha(1-\beta)\big]\,.$

The measure of the set on the right can be estimated with Lemma 10.3: (10.9) follows. ∎

Corollary 10.9 *Let $1 < p < \infty$. If $f \ge 0$ is p–integrable then so is D^*f.*

Proof. We leave it to the reader to check that the hypotheses of the Fubini–Tonelli Theorem 2.2 are satisfied. The argument is a double application of Corollary 2.4 on p. 132, once with $\Phi(t) = t^p$, then again with $\Phi(t) = t^{p-1}$.

$$\int (D^*f)^p\; d\lambda = \int_0^\infty p\alpha^{p-1}\lambda([D^*f > \alpha])\,d\alpha$$

$$\leq \int_0^\infty \frac{p\alpha^{p-1}}{\alpha(1-\beta)} \int_{-\infty}^\infty f(s)[f(s) > \alpha\beta]\, ds\, d\alpha$$
$$= \frac{p}{1-\beta} \int_{-\infty}^\infty \int_0^\infty \alpha^{p-2}[\alpha < g(s)/\beta]\, d\alpha\, ds$$
$$= \frac{p}{(1-\beta)(p-1)} \int_\infty^\infty \Big(g(s)/\beta\Big)^{p-1} g(s)\, ds < \infty\,.$$

Actually, there is an estimate of the $\mathcal{L}^p$–mean of the maximal function D^*f in terms of the $\mathcal{L}^p$–mean of f:

Theorem 10.10 (Hardy–Littlewood) *Let $1 < p < \infty$, and let $f \geq 0$ be a locally Lebesgue integrable function. Then (p' is the conjugate exponent of p)*

$$\|D^*f\|_p \leq p' \cdot \|f\|_p\,.$$

Proof. UsingHölder's inequality and the theorem of Fubini–Tonelli and its consequences, in the same capacity as in the previous proof, we obtain the following chain of inequalities from the Hardy–Littlewood Maximal Lemma:

$$\big(\|D^*f\|_p\big)^p = \int (D^*f)^p\, d\lambda = p\int_0^\infty \alpha^{p-1}\lambda\big(\big[D^*f > \alpha\big]\big)\, d\alpha$$
$$\leq \int_0^\infty p\alpha^{p-1}\Big(\frac{1}{\alpha}\int_{\mathbb{R}} f\big[D^*f > \alpha\big]\, d\lambda\Big)\, d\alpha$$
$$= \int_{\mathbb{R}}\int p\alpha^{p-2}\big[\alpha < (D^*f)(s)\big] f(s)\, d\alpha\, ds$$
$$= \frac{p}{p-1}\int_{\mathbb{R}} (D^*f)^{p-1}(s)\cdot f(s)\, ds \leq \frac{p}{p-1}\|(D^*f)^{p-1}\|_{p'}\cdot \|f\|_p$$
$$= p'\cdot \big(\|D^*f\|_p\big)^{1-\,1/p}\cdot \|f\|_p\,.$$

Since we know already that $\|D^*f\|_p$ is finite we may divide by $\|D^*f\|_p$. The desired inequality follows.

Appendix A

Answers to Selected Problems

I.2.1 "There is a point x in the common domain at which $f(x) > g(x)$."

I.2.4 In both cases take $\Phi(x) = \arctan(x)$.

I.2.5 $x \in [f_n \not\to]$ iff the sequence $(f_n(x))$ of reals is not Cauchy. This in turn means that there is an $r \in \mathbb{N}$ such that for every $p \in \mathbb{N}$ there are $m, n \geq p$ with $|f_m(x) - f_n(x)| > 1/r$.

I.2.12 $P(t) = a_1 x + a_2 x^2 + \cdots + a_n x^n \implies P \circ \phi = a_1 \phi + a_2 \phi^2 + \cdots + a_n \phi^n \in \mathcal{E}$.

I.2.13 Sketch: $|f| = f \vee 0 - f \wedge 0$; $f \wedge g = f + g - (f \vee g)$; $f \vee g = \frac{1}{2}(f + g + |f - g|)$; and $|f| = f_+ + f_- = 2f_+ - f$, where $f_+ \stackrel{\text{def}}{=} f \vee 0$ and $f_- \stackrel{\text{def}}{=} (-f) \vee 0$

I.2.14 (i): $\phi \vee (-r) = -(-\phi \wedge r) = -r(-\frac{1}{r}\phi \wedge 1)$. (ii) The spaces of all or all bounded or all bounded continuous functions on $\mathbb{R}^n$. (iii) The spaces of continuous functions on the line that vanish at infinity or have compact support; the spaces of step functions and of Riemann integrable functions.

I.2.15 Answer for the case that $\mathcal{E}$ is a vector lattice closed under chopping: Let $\phi + r, \phi' + r'$ be two elements of $\mathcal{E} + \mathbb{R}$. It suffices to show that the infimum $\psi = (\phi + r) \wedge (\phi' + r')$ belongs to $\mathcal{E} + \mathbb{R}$ (See I.2.13 (i).) This is clear if $r = r'$, for then $\psi = (\phi \wedge \phi') + r$. If $r' > r$ we write $\psi = (\phi - \phi') \wedge (r' - r) + \phi + r$, which exhibits ψ as an element of $\mathcal{E} + \mathbb{R}$. The case $r < r'$ is handled the same way.

I.2.18 For $0 \leq \phi \in \mathcal{E}$ set $\phi_n = \phi - \phi \wedge 1/n$. Then $\phi_n \leq \phi$ is confined by $n\phi \in \mathcal{E}$ and $\phi - 1/n \leq \phi_n \leq \phi$.

I.3.3 For example, from the Calculus

$$\left| \int f(x) g(x)\, dx \right| \leq \|g\|_u \cdot \int |f(x)|\, dx ;$$

uniform limits of continuous functions are continuous; power series are infinitely often differentiable.

I.3.8 $\|f - f_n\|_u \leq \epsilon \iff |f(x) - f_n(x)| \leq \epsilon \quad \forall\, x$.

I.3.15 (i): The collection $\mathcal{P}$ of polynomials $p(t) = \sum_{k=0}^{K} a_k t^k$ is an algebra separating the points — because the polynomial $p(t) = t$ does on its own — on every compact set. The collection of polynomials has no common zero. (ii): $x \mapsto |x|$ or $x \mapsto e^x$ qualify. (iii): There are polynomials p_n uniformly as close as $1/n$ to ϕ on $[-n, n]$. If $\phi(0) = 0$ they can be chosen so that $p_n(0) = 0$ as well. They meet the description.

I.3.16 (i): By Corollary I.3.12, $\overline{\psi} \stackrel{\text{def}}{=} \sqrt{\overline{\phi}}$ belongs to $\overline{\mathcal{E}}$. Let $\epsilon > 0$ be given and set $M = \|\overline{\psi}\|_u + \epsilon$. There is a $\psi \in \mathcal{E}$ with $\|\overline{\psi} - \psi\|_u < \epsilon/2M$. Set $\phi = \psi^2$. Then $\phi \in \mathcal{E}_+$ and $|\overline{\phi} - \phi| = |\overline{\psi} - \psi| \cdot |\overline{\psi} + \psi| \leq \epsilon/2M \cdot 2M = \epsilon$.
(ii): The condition is clearly necessary. Suppose it is satisfied, and $\phi \in \mathcal{E}$ satisfies $\phi \geq \alpha$ for some $\alpha > 0$. The function $f : t \mapsto 1/t$ is on $[\alpha, \|\phi\|_u]$ the uniform limit of a sequence of polynomials p_n. Evidently $p_n \circ \phi \cdot \phi \in \mathcal{E}$ converges uniformly to $f \circ \phi \cdot \phi = 1$.

K. Bichteler, *Integration – A Functional Approach*, Modern Birkhäuser Classics,
DOI 10.1007/978-3-0348-0055-6_6, © Birkhäuser Verlag 1998

I.3.17 There are polynomials p_n^0 with $\|\psi - p_n^0\|_u \le 4^{-n}$. The polynomials $p_n = p_n^0 - 2^{-n}$ answer. For the second claim let $\phi_n^0 \in \mathcal{E}$ have $\|\overline{\phi} - \phi_n^0\|_u \le 4^{-n}$. Since $\mathcal{E}$ is assumed to contain the constants, the functions $\phi_n^\wedge = \phi_n^0 - 2^{-n}$ and $\phi_n^\vee = \phi_n^0 + 2^{-n}$ belong to $\mathcal{E}$, and $\phi_n^\wedge \uparrow_n \overline{\phi}$ and $\phi_n^\vee \downarrow_n \overline{\phi}$ uniformly.

I.3.18 There are polynomials p_n^0 that increase uniformly on $[0, M]$ to the continuous function $t \mapsto \psi^0(t) = (t \wedge 1)/t$. Since ψ^0 is strictly positive, the p_n^0 will eventually be so as well. The polynomials $p_n(t) \stackrel{\text{def}}{=} tp_n^0(t)$ answer the first claim, the functions $p_n \circ \phi \in \mathcal{E}_+$ the second, with $M = \|\phi\|_u$.

I.3.19 (i): There are an increasing sequence $(p_n^\wedge)$ and a decreasing sequence $(p_n^\vee)$ of polynomials that converge uniformly on $[-M, M]$ to the continuous function $t \mapsto \psi^0(t) = (t - (t \wedge \epsilon))/t^2$. The polynomials $t^2 p_n^\wedge(t)$ and $t^2 p_n^\vee(t)$ answer the first claim, their compositions with ϕ the second, with $M = \|\phi\|_u$.
(ii): For $\phi \in \mathcal{E}_+$, both $(\phi - \epsilon)_+$ and $(-\phi - \epsilon)_+$ are confined by $\phi^2/\epsilon^2 \in \mathcal{E}$ and their sum differs from ϕ by less than ϵ. The uniform closure of $\overline{\mathcal{E}}_{00}$ therefore contains $\mathcal{E}$ and then $\overline{\mathcal{E}}$. The $\mathcal{E}$–confined bounded functions form both a ring and a vector lattice closed under chopping, and then so does their intersection with $\overline{\mathcal{E}}$, $\overline{\mathcal{E}}_{00}$. If some bounded f is confined by $\phi \in \mathcal{E}$ then it vanishes on $\phi \ge 1$ and also on the set $[(2\phi - 1)_+ \ge 1]$, showing that it is $\overline{\mathcal{E}}_{00}$–confined.

I.3.20 It suffices to treat the case $0 < \alpha < 1$ — for $\alpha > 1$ write t^α as the product of the polynomial $t^{[\alpha]}$ with $t^{\alpha - [\alpha]}$.
There is no loss in assuming that $M \ge 1$. Set $\delta = \big(1 \wedge \epsilon \wedge (\alpha M^{\alpha-1})\big)/4M$. There is a polynomial $P'(s)$ which differs from the continuous function

$$s \mapsto \alpha s^{\alpha-1} \wedge \delta^{(\alpha-1)/\alpha}$$

by less than δ, uniformly on $[0, M]$. The desired polynomial is

$$P(t) = \int_0^t (P'(s) - \delta)\, ds\,.$$

Since $0 \le P'(s) - \delta \le \alpha s^{\alpha-1} \wedge \delta^{(\alpha-1)/\alpha} \le \alpha s^{\alpha-1}$, we have $0 \le P(t) \le t^\alpha$. On the other hand,

$$\begin{aligned} t^\alpha - P(t) &= \int_0^t \Big(\alpha s^{\alpha-1} - P'(s) + \delta\Big)\, ds \\ &= \int_0^t \Big((\alpha s^{\alpha-1} \wedge \delta^{(\alpha-1)/\alpha}) - P'(s) + \delta\Big)\, ds \\ &+ \int_0^t \Big(\alpha s^{\alpha-1} - (\alpha s^{\alpha-1} \wedge \delta^{(\alpha-1)/\alpha})\Big)\, ds\,. \end{aligned}$$

The first integral is less than $2\delta M \le \epsilon/2$. The integrand of the second vanishes for $s > \delta^{1/\alpha}$, so the integral itself is less than $\int_0^{\delta^{1/\alpha}} \alpha s^{\alpha-1}\, ds = \delta \le \epsilon/2$.

I.3.21 Let $\psi \in C_0[\mathbb{R}]$. There is an extension of ψ to a continuous function $\widehat{\psi}$ on the compact space $\overline{\mathbb{R}}_+ = [0, \infty]$ (Exercise I.2.4), gotten by setting $\widehat{\psi}(\infty) = 0$. The functions of the form

$$\phi(t) = \sum_{i=1}^{I} r_i \cdot e^{\alpha_i t} \qquad \alpha_i > 0\,,$$

form a ring $\mathcal{E}$ and have a similar extension $\widehat{\phi}$ each. The $\widehat{\phi}$ form a ring $\widehat{\mathcal{E}}$ of continuous functions on the compact space $\overline{\mathbb{R}}_+$ that separates the points, and the common zero is $z = \infty$. Now apply Theorem I.3.11. For the second statement observe that $\mathcal{L}f = 0$ implies $\int f(t)e^{\alpha t}\phi(t)\,dt = 0$ for all $\alpha > 0$ and all $\phi \in \mathcal{E}$ and then for all $\phi \in \overline{\mathcal{E}}$.

I.3.22 Hint: The complex conjugate of f is the function $\overline{f} : t \mapsto \overline{f(t)}$, its real part is the function $\Re f : t \mapsto \frac{1}{2}(f + \overline{f})$, and its imaginary part is the real–valued function $\Im f = \frac{1}{2i}(f - \overline{f})$, so that $f = \Re f + i\Im f$. The real–valued functions in $\mathcal{E}$ form a ring $\Re\mathcal{E}$ separating the points. Given any continuous function that vanishes on the common zero of $\mathcal{E}$ — if there is one — approximate its real and imaginary parts separately by functions in $\Re\mathcal{E}$.

I.3.27 Let $\widehat{f}$ be the extension of f to a function from $\overline{S}$ to the completion $\overline{E}$ of E. Its range is compact and contains the range of f.

I.3.28 If $\overline{\mathcal{E}}$ does not contain the constants then the functions $\overline{\phi} \in \overline{\mathcal{E}}$ have a common zero z in the Hausdorff completion $\overline{S}$. In that case let $\widehat{S} = \overline{S} \setminus \{z\}$, else set $\widehat{S} = \overline{S}$. In both cases $\widehat{S}$ is locally compact. For every $\overline{\phi} \in \overline{\mathcal{E}}$ let us denote by $\widehat{\phi}$ the unique function on $\widehat{S}$ with $\widehat{\phi} \circ j = \overline{\phi}$. If $0 \le \overline{\phi} \in \overline{\mathcal{E}}_{00}$ then $\widehat{\phi}$ is a function of compact support. There are a $\sigma \in \mathcal{E}$ with $\widehat{\sigma} \ge [\widehat{\phi} > 0]$, by Tietze's extension theorem a function $\overline{\rho} \in C_{00}[\widehat{S}]$ which on $[\widehat{\sigma} \ge 1]$ equals $1/\widehat{\sigma}$, a sequence $\rho_n \in \mathcal{E}$ converging uniformly to $\overline{\rho}$, and a sequence $\phi'_n \in \mathcal{E}$ converging uniformly to $\overline{\phi}$. Then $\phi_n = \phi'_n \cdot \sigma \cdot \rho_n \in \mathcal{E}$ is confined by any function ψ confining σ, and $\phi_n \to \overline{\phi}$ uniformly.

I.4.6 (i) Let $\phi \in \mathcal{E}[\mathbb{R}]$ be a set; i.e., ϕ takes only the values 0 or 1. Let $\mathcal{P} = \{x_0 < \cdots < x_n\}$ be a partition for ϕ. Clearly ϕ is the union of those intervals $(x_{i-1}, x_i]$ on which ϕ equals 1. Let I denote the minimal number of half–open intervals needed to represent A, and let $A = \bigcup_{i=1}^{I}(a_i, b_i]$ be a representation consisting of I intervals. Clearly $(a_i, b_i]$ and $(a_j, b_j]$ cannot overlap or be adjacent for $1 \le i \ne j \le I$; else we would remove them from the representation of A and add their union, which is another half–open interval, arriving at a representation with less than I intervals. The intervals of our representation are thus disjoint, even non–adjacent. The uniqueness of the minimal representation is left to the reader.
(ii): The sets in *any* vector lattice $\mathcal{E}$ of functions form a ring of sets: if $A, B \in \mathcal{E}$ then $A \cap B = A \wedge B \in \mathcal{E}$, etc.
Here is another, if more complicated, way of seeing this: Suppose $A = \bigcup_{i=1}^{I}(a_i, b_i]$ and $A' = \bigcup_{i=1}^{I'}(a'_i, b'_i]$ belong to $\mathcal{A}[\mathbb{R}]$. It is evident that $A \cup A'$ is the union of not more than $I + I'$ half–open intervals. $A \cap A'$ is the union of the half–open intervals $(a_i, b_i] \cap (a'_j, b_j]$, $1 \le i \le I, 1 \le j \le I'$, finite in number. Finally, $(a_i, b_i] \setminus (a'_j, b'_j]$ is the union of at most two half–open intervals, so $(a_i, b_i] \setminus A'$ is the finite union of half–open intervals, and then so is their union over $i \in I$, to wit, $A \setminus A'$.
(iii): these are the definitions of $\mathcal{E}[\mathbb{R}]$ and the elementary integral.

I.5.1 $\int^\natural f \le \int^\natural (f - f') + \int^\natural f' \le \int^\natural |f - f'| + \int^\natural f'$ implies $\int^\natural f - \int^\natural f' \le \int^\natural |f - f'|$. Reversing the rôles of f and f' yields Inequality (I.5.2).

I.6.1 For the first questions invoke the permanence properties for continuity and differentiability, which are, for continuity: algebraic and order–combinations of continuous functions are continuous; the uniform limit of continuous functions is continuous. For differentiability they are: sum rule, Leibniz' rule, quotient rule, chain rule, inverse function rule, and a limit theorem too involved to spell out here.

I.6.6 The sequence $(-n, n]/n$ converges uniformly to the dominated function 0. Count the rationals in the unit interval I: $\{q_1, q_2, \ldots\}$, and let f_n be the indicator function of $\{q_1, q_2, \ldots, q_n\}$. Everyone of them is Riemann integrable with integral 0. (f_n) converges dominatedly to (the indicator function of) the rationals in I, which is not Riemann integrable.

I.6.7 The support of a function f is the closure $\overline{[f \neq 0]}$ of its carrier $[f \neq 0]$. Since the support of f is assumed compact, there is an interval $(-M, M)$ outside which f vanishes. Given $n \in \mathbb{N}$ there is a $\delta_n > 0$ such that $|s-t| < \delta_n$ implies $|f(s) - f(t)| < 1/n$, since f is uniformly continuous on $[-M, M]$. Let ϕ_n be the function that equals $f(k\delta_n)$ on the interval $((k-1)\delta_n, k\delta_n]$. Evidently ϕ_n is a step function and differs from f uniformly by less than $1/n$. The sequence (ϕ_n) is majorized by a suitable multiple of the step function $(-M, M]$ and converges uniformly to f as $n \to \infty$. Its limit f is therefore integrable.

I.6.8 Let $\epsilon > 0$ be given, set $M = \|f\|_u + \|f'\|_u$, and let $\{a_1, \ldots, a_n\}$ be the points at which f differs from f'. Set

$$\psi = \sum_{i=1}^{n} |f(a_i) - f'(a_i)| \cdot \left(a_i - \epsilon 2^{-i-2}/M, a_i + \epsilon 2^{-1-2}/M\right] .$$

Then $|f - f'| \leq M\psi$ and thus $\|f - f'\|^{\natural} \leq \int M\psi < \epsilon/2$. Now find $\phi \in \mathcal{E}$ with $\|f - \phi\|^{\natural} < \epsilon/2$ and use the subadditivity of $\| \ \|^{\natural}$ to conclude that $\|f' - \phi\|^{\natural} < \epsilon$.

I.6.9 A function f is ***piecewise continuous*** if there is a partition $-\infty < x_0 < x_1 < \cdots < x_n < +\infty$ such that f is continuous on every one of the intervals (x_{i-1}, x_i) and has a limit at its endpoints. By Exercise I.6.12 $f_i = f \cdot (x_{i-1}, x_i)$ is integrable, and f differs from $\sum f_i$ in at most finitely many points. Thus f is integrable (Exercise I.6.8).

I.6.10 This is not easy to show using the original definition of integrability. You might try. Also, there is no easy way to express the integral of the infimum or product in terms of the integrals of the two functions. This renders the information useless for computations of the integral.

I.6.11 If ϕ is a polynomial with zero constant term then this is clear, as $\mathcal{L}^{\natural}$ is a ring. By Theorem I.3.11, an arbitrary ϕ can be approximated uniformly by such polynomials p_n, and as the $p_n \circ f$ vanish on some fixed interval containing the carrier of f, its uniform limit $\phi \circ f$ is Riemann integrable.

I.6.12 Let $f_{\leftarrow}(a) = \lim_{s \downarrow a} f(s)$ and $f_{\leftarrow}(x) = f(x)$ everywhere else. Next let $\widehat{f}$ be the function equal to $f_{\leftarrow}$ on $[a, b]$, zero on $(-\infty, a-1]$ and on $[b+1, \infty)$, and linear in the remaining two intervals $[a-1, a]$ and $[b, b+1]$. This function is continuous on the whole line and has compact support $\subset [a-1, b+1]$. It is therefore integrable (Corollary I.6.7). Its product $f = \widehat{f} \cdot 1_{(a,b]}$ is therefore integrable as well (Section I.6).

I.6.13 (i): $x \to 1/x$ is continuous on $(0, 1]$ without being integrable.
(ii): $|F(b') - F(b) - f(b)(b'-b)| \leq |\int_b^{b'} f(x)\,dx - \int_b^{b'} f(b)\,dx| = |\int_b^{b'} (f(x) - f(b))\,dx| \leq |b'-b| \cdot \sup\{|f(x) - f(b)| : x \text{ between } b \text{ and } b'\} = o(b - b')$.

I.6.16 Set

$$\underline{f}(x) = \lim_{\delta \to 0} \inf_{|\xi - x| < \delta} f(\xi) \quad \text{and} \quad \overline{f}(x) = \lim_{\delta \to 0} \sup_{|\xi - x| < \delta} f(\xi) ,$$

and let $\underline{\phi}, \overline{\phi}$ be the continuous functions of compact support provided by Exercise I.6.14. Then clearly

$$\underline{\phi} \leq \underline{f} \leq f \leq \overline{f} \leq \overline{\phi} .$$

Since $\int(\overline{\phi} - \underline{\phi}) < \epsilon$ and $\epsilon > 0$ is arbitrary, the open(!) set

$$[\overline{f} - \underline{f} > 1/n] \subset [\overline{\phi} - \underline{\phi} > 1/n] \leq n(\overline{\phi} - \underline{\phi})$$

is negligible. So is therefore the set

$$[\overline{f} - \underline{f} > 0] = \bigcup_n [\overline{f} - \underline{f} > 1/n]$$

of discontinuities of f. This shows the condition to be necessary. Hint for the sufficiency: let $[a,b]$ be a compact interval outside which f vanishes, and let $\epsilon > 0$. Find open intervals (a_i, b_i) with $\sum(b_i - a_i) < \epsilon$ whose union contains the set of discontinuity points of f. for every continuity point x of f find an open interval I_x about x on which f varies less than ϵ. Extract a finite subcover and use it to construct suitable step functions below and above f.

I.7.3 Let $f_n \to f$. If $\|f - g\| = 0$ then $f_n \to g$ since $\|f_n - g\| \le \|f_n - f\| + \|f - g\| \to 0$. Conversely, if $f_n \to g$ as well, then $\|g - f\| \le \|g - f_n\| + \|f_n - f\| \to 0$ and $\|g - f\| = 0$. If $\| \ \|$ is a norm, $\|f - g\| = 0 \implies f = g$.

I.7.4 Assume the subsequence (f_{n_k}) of the sequence (f_n) has limit f. Let $\epsilon > 0$ be given. There is an $N \in \mathbb{N}$ such that $\|f_m - f_n\| < \epsilon/2$ for $m, n \ge N$. There is a $K \in \mathbb{N}, K \ge N$ such that $\|f - f_{n_k}\| < \epsilon/2$ for $k \ge K$. For $n \ge n_K \, (\ge N)$, $\|f - f_n\| \le \|f - f_{n_K}\| + \|f_{n_K} - f_n\| \le \epsilon/2 + \epsilon/2 = \epsilon$.

I.7.6 The norm of a class $f + \mathcal{N} \in L$ is defined by $\|f + \mathcal{N}\| = \|f\|$. The contractivity of the seminorm established in Exercise I.7.2 shows that this number does not depend on the representative f chosen.

I.7.7 Sketch: A sequence (f_n) of functions converges to f in the sense of the seminorm $\| \ \|_u$ if it converges to f uniformly. Thus if the f_n are continuous so is f. If the f_n are bounded or vanish at infinity then so does f.
To see that $(C_{00}^{\infty}[\mathbb{R}], \| \ \|_u)$ is not complete, prove that the function ϕ which is zero for $t \le 0$ and has the value $\phi(t) = \exp[-1/t^2]$ for $t > 0$ is infinitely often differentiable. Set $\psi(t) = \phi(1+t) \cdot \phi(1-t)$ and $\mathcal{E} = \{\psi \cdot p : p \text{ a polynomial }\} \subset C_{00}^{\infty}[\mathbb{R}]$. The function $t \mapsto (1/2 - |t|) \vee 0$ can be uniformly approximated by elements of $\mathcal{E}$ (why?) yet is not differentiable.

I.7.8 (i) The function $f = \{0\}$ which equals 1 at 0 and zero elsewhere has $\|f\|^{\natural} = 0$ without being zero itself. (ii) Let $0 < q < 1$. Set $C_0 = [0,1]$. This interval has length $\ell_0 = 1$. Let C_1 be the set obtained by removing from C_0 the interval of length $q\ell_0$ that is centered at the midpoint of C_0. C_1 consists of two intervals that each have length $\ell_1 = (1-q)\ell_0/2$. From each of them remove the interval of length $q\ell_1$ that is centered at its midpoint, obtaining the set C_2. Continue on. The sequence C_n decreases to a compact set C, which is called ***a Cantor set.*** (For $q = 1/3$ it is called ***the Cantor set.***) The sequence (C_n) is Cauchy in $\| \ \|^{\natural}$–mean, yet has no Riemann integrable limit in $\| \ \|^{\natural}$–mean.

I.7.10 Hint: see the proof of Theorem IV.1.2. (i): $a^{\lambda} \cdot b^{1-\lambda} \le \lambda \cdot a + (1-\lambda) \cdot b$ for $a, b \ge 0$ and $0 \le \lambda \le 1$. (ii): Using this with $\lambda = 1/p, 1 - \lambda = 1/p'$, $a = |x_i|^p / \|x\|_p^p$, $b = |y_i|^{p'} / \|y\|_{p'}^{p'}$, prove Hölder's inequality

$$\sum_{i=1}^{n} |x_i y_i| \le \|x\|_p \cdot \|y\|_{p'} \, .$$

(iii): Write $|x_i + y_i|^p \le |x_i + y_i|^{p-1}(|x_i| + |y_i|)$ and apply Hölder's inequality with suitable exponents to obtain the subadditivity of $\| \ \|_p$.

I.7.14 Let $\mathcal{D}_n = \{(q_1, q_2, \ldots, q_n, 0, 0, \ldots) : q_i \in \mathbb{Q}\}$. Then $\mathcal{D} = \bigcup_{n=1}^{\infty} \mathcal{D}_n$ is countable and dense if $1 \le p < \infty$. To show that ℓ^{∞} is not separable consider the ***power set*** of $\mathbb{N}$, the set $\mathfrak{P}[\mathbb{N}]$ of all subsets of $\mathbb{N}$. It contains uncountably many functions (Convention I.2.6) any two of which have distance 1 from each other. Given a dense subset $\mathcal{D}$ of ℓ^{∞}, construct an injection $\mathfrak{P}[\mathbb{N}] \to \mathcal{D}$ by picking for any $A \in \mathfrak{P}[\mathbb{N}]$ an element $d_A \in \mathcal{D}$ whose distance from A is less than $1/3$: $\mathcal{D}$ cannot be countable.

I.7.15 Let $\mathcal{E}$ denote the collection of functions f of the form

$$f(t) = \sum_{n=1}^{N} e^{-q_n t^2} \cdot p_n(t) \ ,$$

where $q_n > 0$ are rational numbers and the p_n are polynomials with rational coefficients. This collection is countable (why?), and its uniform closure $\overline{\mathcal{E}}$ is a ring separating the points of $\mathbb{R}$ (why?). Despite the fact that $\mathbb{R}$ is not compact you can show with Weierstraß's theorem that $\overline{\mathcal{E}} = C_0[\mathbb{R}]$. This shows that $C_0[\mathbb{R}]$ contains $C_0^\infty[\mathbb{R}]$ densely.

I.7.17 Let $f \in \overline{\mathcal{E}}$. There exists a sequence (ϕ_n) in $\mathcal{E}$ with $\|\phi_n - f\| \to 0$. If $\overline{I}$ is to be majorized by $\| \ \|$, then $|\overline{I}(f) - I(\phi_n)| = |\overline{I}(f) - \overline{I}(\phi_n)| = |\overline{I}(f - \phi_n)| \le \|f - \phi_n\| \to 0$, and we must define $\overline{I}(f) = \lim I(\phi_n)$. Thus, if the extension exists, it is unique. Let us establish that $(I(\phi_n))$ does, indeed, have a limit. By the subadditivity of $\| \ \|$, $|I(\phi_m) - I(\phi_n)| = |I(\phi_m - \phi_n)| \le \|\phi_m - \phi_n\| \le \|\phi_m - f\| + \|f - \phi_n\| \to 0$ as $m, n \to \infty$. This shows that $(I(\phi_n))$ is Cauchy and thus has a limit, inasmuch as $(\mathbb{R}, | \ |)$ – or $(\mathcal{F}', \| \ \|')$, respectively – is complete. This limit is unique, since the range space is normed and not merely seminormed – see Exercise I.7.3. If (d'_n) is a second sequence in $\mathcal{E}$ converging to f then $\|d_n - d'_n\| \to 0$ from the subadditivity of $\| \ \|$ again, hence $|I(d_n) - I(d'_n)| \to 0$, and the limit is the same. The extension $\overline{I}$ is thus well–defined. Its linearity is now easy to check: If $f, f' \in \overline{\mathcal{E}}$ are limits of the sequences (ϕ_n), (ϕ'_n), respectively, then $rf + r'f'$ is the limit of $(r\phi_n + r'\phi'_n)$, since

$$\|(rf + r'f') - (r\phi_n + r'\phi'_n)\| \le |r| \cdot \|f - \phi_n\| + |r'| \cdot \|f' - \phi'_n\| \to 0 \ .$$

Consequently

$$\overline{I}(rf + r'f') = \lim I(r\phi_n + r'\phi'_n) = \lim rI(\phi_n) + r'I(\phi'_n) = r\overline{I}(f) + r'\overline{I}(f') \ .$$

I.7.20 See Exercises I.3.22 and I.3.21.

I.7.21 Define the upper and lower integrals of a function f on S by

$$\int^{\natural} f = \inf\{\int \phi : \ f \le \phi \in \mathcal{E}\} \quad \text{and}$$

$$\int_{\natural} f = \sup\{\int \phi : \ f \ge \phi \in \mathcal{E}\}, \ \text{respectively.}$$

$\int^{\natural} f$ may be infinite, namely $\int^{\natural} f = -\infty$ if $\{\int \phi : f \le \phi \in \mathcal{E}\}$ is not bounded below or $\int^{\natural} f = +\infty$ if there is no elementary function ϕ (i.e., $\phi \in \mathcal{E}$) majorizing f. Define $\mathcal{F}^{\natural} = \{f : \|f\|^{\natural} \stackrel{\text{def}}{=} \int^{\natural} |f| < \infty\}$, let $\mathcal{L}^{\natural}$ denote the closure of $\mathcal{E}$ in $\mathcal{F}^{\natural}$ and extend the integral by continuity, for example. The theorems and their proofs can now be copied word–for–word: Nowhere have we used the structure of the underlying set nor that the elementary integral of a step function is the sum *height–of–step* × *step–size*.

II.2.5 Let $\| \ \|$ denote the usual euclidean distance on $\mathbb{R}^n$. The function

$$s \mapsto d(s) = \inf_{k \in K} \|s - k\| \ , \qquad s \in S,$$

is continuous on S (why?) and vanishes on K. The function $\psi = (1 - d)_+$ meets the description.

II.2.7 Again by Dini's theorem, Φ converges to its pointwise supremum ψ uniformly in the sense that for every $\epsilon > 0$ there exists an $\phi_\epsilon \in \Phi$ such that $\Phi \ni \phi \geq \phi_\epsilon \implies \|(\psi - \phi)\|_u < \epsilon \implies \psi - \phi_\epsilon \leq \epsilon\rho \implies |m(\psi) - m(\phi_\epsilon)| \leq \epsilon m(\rho)$ for some $\rho \in C_{00}$ confining ψ.

II.2.12 (iii): Let $\mathcal{C}_0^1 = \mathcal{C}_0^2 = \mathcal{C}$. If $\mathcal{C}_n^1 = \mathcal{C}_n^2 = \mathcal{C}$ have been defined let $\mathcal{C}_{n+1}^1$ denote the collection of sets that are finite unions of sets in $\mathcal{C}_n^2$, and let $\mathcal{C}_{n+1}^2$ be the collection of sets that are relative complements of two sets in $\mathcal{C}_{n+1}^1$. Then $[\mathcal{C}] = \bigcup_{n=1}^\infty \mathcal{C}_n^2$ (Why?). Thus if $\mathcal{C}$ is countable so is $[\mathcal{C}]$. For the case of finite $\mathcal{C}$ use Lemma II.2.14.

II.3.7 $[f > r] \leq |f|/r$. Apply $\|\ \|^*$ and use its solidity.

II.3.10 This is rather easy to prove if one develops the integration theory of $\|\ \|^*$ first (see Proposition III.6.2 (iv)). Here is a direct proof: Since $\int^*$ is increasing, the left–hand side is certainly not smaller than the right–hand side. Only the converse inequality

$$\int^* f \leq \sup \int^* f_n \tag{?}$$

needs to be proved; and it is certainly satisfied if the supremum on the right is infinite. Suppose then that $\int^* f_n < \infty$ for all n, and let $\epsilon > 0$ be given. For each n there is a function $h_n \in \mathcal{E}^\uparrow$ majorizing f_n and so that

$$\int^* h_n < \int^* f_n + \epsilon 2^{-n}.$$

Let $h_n' = \bigvee_{k \leq n} h_k$. Next prove by induction the following claim

$$\int^* h_n' \leq \int^* f_n + \epsilon(1 - 2^{-n}) \leq \int^* f_n + \epsilon\,. \tag{$*$}$$

Suppose this has been shown. Then, as $f \leq h' = \bigvee h_n' = \bigvee h_n \in \mathcal{E}^\uparrow$,

$$\int^* f \leq \int^* h' = \sup_n \int^* h_n' \leq \sup_n \int^* f_n + \epsilon,$$

and the desired inequality (?) follows, since $\epsilon > 0$ was arbitrary.
Now to the proof of $(*)$. It is certainly true if $n = 1$:

$$\int^* h_1' = \int^* h_1 \leq \int^* f_1 + \epsilon 2^{-1} = \int^* f_1 + \epsilon(1 - 2^{-1}).$$

Suppose $(*)$ is true for n. Using the additivity of $\int^*$ on $\mathcal{E}^\uparrow$ we write the following chain of (in)equalities, which yield the claim for $n + 1$:

$$\begin{aligned}
h_{n+1}' + h_n' \wedge h_{n+1} &= h_n' \vee h_{n+1} + h_n' \wedge h_{n+1} \\
&= h_n' + h_{n+1} \\
\implies \quad \int^* h_{n+1}' + \int^* h_n' \wedge h_{n+1} &= \int^* h_n' + \int^* h_{n+1} \\
\implies \quad \int^* h_{n+1}' - \int^* h_{n+1} &= \int^* h_n' - \int^* h_n' \wedge h_{n+1} \\
&\leq \int^* h_n' - \int^* f_n \leq \epsilon(1 - 2^{-n}) \\
\implies \quad \int^* h_{n+1}' - \int^* f_{n+1} &\leq \int^* h_{n+1}' - \int^* h_{n+1} + \int^* h_{n+1} - \int^* f_{n+1} \\
&\leq \epsilon(1 - 2^{-n}) + \epsilon 2^{-(n+1)} = \epsilon(1 - 2^{-(n+1)}) < \epsilon\,.
\end{aligned}$$

II.3.11 See the proof of Theorem IV.1.6.

II.3.12 It is impossible to prove this as it is false, in general. If the f_n are integrable and $\int^* f$ is finite then it is true (see Theorem II.5.4). It is hard to disprove the general statement of the exercise. An example of a sequence (f_n) decreasing pointwise to zero but with $\int^* f_n \not\to 0$ is produced in Exercise II.8.1 on p. 67.

II.3.14 After (i) do (iii): it is merely a rewrite of the definition. (ii) follows immediately from the corresponding results about $\int^*$. (iv): If $\mathcal{E}_\downarrow \ni k \le f \le h \in \mathcal{E}^\uparrow$ then $h - k \in \mathcal{E}^\uparrow_+$ and $\int^* h - \int_* k = \int^*(h-k) \ge 0$.

II.3.15 Let (ϕ_n) be sequence of positive step functions increasing to the function 1. By the continuity along increasing sequences of Exercise II.3.10, $\int^* f = \sup \int^* \phi_n \cdot f \le \sup\{\cdots\} \le \int^* f$.

II.4.2 (iii): Note first that $\|(a,b)\|^* = b - a$ for an open interval (a,b), since $(a,b) = \sup_n (a, b - \frac{1}{n}] \in \mathcal{E}^\uparrow$. Next enumerate the rationals: $\mathbb{Q} = \{q_1, q_2, \ldots\}$. The open set $U = \bigcup_{n=1}^\infty (q_n - \epsilon 2^{-n}, q_n + \epsilon 2^{-n})$ belongs to $\mathcal{E}^\uparrow$ and has measure $\|U\|^* \le 2\epsilon$.
(ii) follows from this, as $\mathbb{Q} \subset U \in \mathcal{E}^\uparrow$ and $\epsilon > 0$ is arbitrary. Another argument is that a one–point–set is clearly Lebesgue negligible, and by countable subadditivity so is $\mathbb{Q}$.

II.4.6 Repeat the calculations of Theorem I.6.3.

II.5.7 For every n let f'_n be a function that is everywhere defined and a.e. equal to f_n where the latter is defined. Then apply the MCT to $f''_n = \bigvee_{k=1}^n f'_k$ if (f_n) is increasing, to $f''_n = \bigwedge_{k=1}^n f'_k$ if (f_n) is decreasing.

II.5.8 Find $\psi_n \in \mathcal{E}$ with $\|f - \psi_n\|^* < 2^{-n}$ and set $\phi_1 = \psi_1$ and $\phi_n = \psi_n - \psi_{n-1}$ for $n \ge 2$.

II.5.9 Suppose h is the pointwise supremum of the sequence (ϕ_k) in the ring $\mathcal{E}$. Then it is the increasing limit of the sequence $\psi_k \stackrel{\text{def}}{=} \phi_1 \vee \ldots \vee \phi_k \in \mathcal{L}^1$. Now if $\sup_k \|\psi_k\|^* < \infty$ then $\phi_k \to h$ in mean and thus h is integrable and $\|\psi_k\|^* \to \|h\|^* < \infty$. Else $\|h\|^* \ge \sup \|\psi_k\|^* = \infty$ and h is not integrable. Next let (h_n) be an increasing sequence in $\mathcal{E}^\uparrow$ with pointwise limit h—which then also belongs to $\mathcal{E}^\uparrow$. Now if $(\|h_n\|^*)$ is bounded then $h_n \to h$ in mean and so $\|h_n\|^* \to \|h\|^*$. Otherwise $\|h\|^* \ge \sup_n \|h_n\|^* = \infty$. In either case $\|h_n\|^* \to \|h\|^*$.

II.5.13 By Theorem II.5.4, $g_N \stackrel{\text{def}}{=} \bigwedge_{n=N}^\infty f_n$ is integrable. Now

$$\|g_N\|^* \le \inf_{n \ge N} \|f_n\|^* \le \liminf_{n\to\infty} \|f_n\|^* .$$

If this last number is infinite then there is nothing to prove. If it is finite, then g_N, which increases pointwise to $g = \liminf f_n$, converges to g in mean by Theorem II.5.4, and

$$\|g\|^* = \lim_n \|g_N\|^* = \liminf_n \|f_n\|^* .$$

In this case, $g = \liminf f_n$ is integrable.

II.5.14 This is done easiest using the notion of mesurability—see Chapter III.

II.5.15 Let $M = \|f\|_u$. Since g is clearly continuous, there is a sequence (p_n) of polynomials with zero constant term converging uniformly on $[-M, M]$ to g. The sequence $p_n \circ f \in \mathcal{L}^1$ converges uniformly to $g \circ f$ and is dominated by $K \cdot |f| \in \mathcal{L}^1$.

II.5.20 Let $\epsilon > 0$ be given. If (f_n) converges in mean then it is mean–Cauchy and one can find $N \in \mathbb{N}$ such that $\|f_N - f_n\|^* < \epsilon$ for $n \ge N$. Set $g = \bigvee_{i=1}^N |f_i| \in \mathcal{L}^1$. It is clear that the distance of f_n from $[-g, g]$ is less than ϵ for all $n \in \mathbb{N}$. The sequence (f_n) is therefore uniformly integrable.

For the converse, we find $g \in \mathcal{L}^1$ so that $f'_n = -g \wedge f_n \vee g$ differs from f_n by less than ϵ in mean, for all $n \in \mathbb{N}$. The sequence (f'_n) is clearly dominated and converges pointwise. It is therefore Cauchy in mean. Thus there is an $N \in \mathbb{N}$ with $\|f'_n - f'_m\|^* < \epsilon/3$ for $n, m \geq N$. The subadditivity of $\| \ \|^*$ yields $\|f_n - f_m\|^* \leq \|f_n - f'_n\|^* + \|f'_n - f'_m\|^* + \|f'_m - f_m\|^* < \epsilon$ for such m, n, showing that (f_n) is mean–Cauchy. The claim follows from Theorem II.5.2.

II.6.2 Assume f integrable. There are $\phi_n \in \mathcal{E}$ with $\|f - \phi_n\|^* = \int^* |f - \phi_n| < 1/n$. Then there are $h_n \in \mathcal{E}^\uparrow$ with $|f - \phi_n| \leq h_n$ and $\int^* h_n < 1/n$. That is to say,

$$\phi_n - h_n \leq f \leq \phi + h_n \quad \text{and}$$

$$\int \phi_n - \int^* h_n = \int_* \phi_n + \int_* -h_n = \int_* (\phi_n - h_n) \leq \int_* f \leq \int^* f \leq \int \phi_n + \int^* h_n .$$

Looking at the extremes we see that the numbers $\int \phi_n$, $\int_* f$, and $\int^* f$ differ by less than $2/n$, which proves Equation (II.6.1).

Conversely, assume that Equation (II.6.1) is satisfied, and let $\epsilon{>}0$. There are $k \in \mathcal{E}_\downarrow$ and $h \in \mathcal{E}^\uparrow$ with $k \leq f \leq h$ and $\int^* h - \int_* k < \epsilon$. There is a $\phi \in \mathcal{E}$ with $\phi \geq k$ and $\int \phi - \int_* k < \epsilon$. Now $0 \leq f - k \leq h - k \in \mathcal{E}^\uparrow$ and thus

$$\int^* |f - k| \leq \int^* h + \int^* (-k) = \int^* h - \int_* k < \epsilon,$$

$$\int^* |\phi - k| = \int^* \phi + \int^* (-k) = \int \phi - \int_* k < \epsilon.$$

By the triangle inequality, $\|f - \phi\|^* = \int^* |f - \phi| < 2\epsilon$. This shows that f can be approximated arbitrarily closely in mean by elementary functions ϕ.

II.6.3 $\int g + \int^*(-f) = \int^*(g - f) \leq \liminf(\int g_n - \int f_n) = \int g - \limsup \int f_n$ implies $\int_* f \geq \limsup \int f_n$ (see II.3.14, II.5.13, and II.6.7). Applying this to the sequence $(-f_n)$ yields $\int^* f \leq \liminf \int f_n$. By Exercise II.6.2, f is integrable and $\int f = \lim \int f_n$. The sequence $(f - f_n)$ satisfies $|f - f_n| \leq 2g_n$ and converges a.e. to zero. Therefore $0 = \lim \int |f - f_n| = \lim \|f - f_n\|^*$. Note that this argument does not apply to arbitrary means, only to the Daniell mean.

II.6.5 There is, by the very definition of $\int^*$, an $h \in \mathcal{E}^\uparrow$ with $h \geq |f|$ and $\int^* h < \infty$. h is integrable as it is the limit of an increasing sequence of elementary functions.

II.6.6 There are functions $h_n \in \mathcal{E}^\uparrow$ with $\int^* h_n \downarrow \int^* f$. Each of them, being the pointwise supremum of an increasing sequence of integrable functions (See the proof of Lemma II.3.1) is integrable as long as its upper integral is finite (MCT). For $\overline{f}$ now take $\lim_{N \to \infty} \bigwedge_{n=1}^N h_n$. To produce $\underline{f}$ use Exercise II.3.14. For the last claim, if f is a set and $\overline{f}$ an upper envelope then $[\overline{f} \geq 1] = \lim_n (\overline{f} \wedge 1)^n$ is a smaller upper envelope. If $\underline{f} \geq 0$ is a lower envelope then $[\underline{f} > 0] = \lim_n 1 \wedge (n\underline{f})$ is a larger lower envelope. A similar proof is given in Proposition III.6.2 on p. 93.

II.6.7 As $\int^*$ is subadditive only the inequality $\int^*(f + f') \geq \int^* f + \int^* f'$ needs proving. Assume f is the integrable function. If $\int^*(f + f') < r \in \mathbb{R}$ then there is a function $h \in \mathcal{E}^\uparrow$ exceeding $f + f'$ and having upper integral $\int^* h < r$. By the MCT, h is integrable. Then $f' \leq h - f \in \mathcal{L}^1$ and consequently $\int^* f' \leq \int^* h - f = \int (h - f) = \int h - \int f$, which implies $\int^* f' + \int^* f \leq r$. The claim follows from of Exercise I.2.3.

II.6.8 Read Section III.6.

II.6.9 Set $\|f\| = \inf\{\int f' : |f| \leq f' \in \mathcal{L}^*\}$.

II.6.10 $C_{00}^{\infty}[\mathbb{R}]$ is dense in $\mathcal{L}^{\natural}$ (Exercise I.7.16). For the second claim use Exercise I.7.15.

II.6.12 Let $\epsilon > 0$ and find $\phi \in C_{00}$ with $\|f - \phi\|^* < \epsilon$. For any function $\psi \in C_{00}$ with $|\psi| \le 1$, $|\int \phi\psi| = |\int (f - \phi)\psi| \le \|f - \phi\|^* < \epsilon$. Apply this to the functions $\psi_n = -1 \vee n \cdot \phi \wedge 1$, which converge pontwise to $\operatorname{sgn}(\phi)$ — in fact, $(\phi \cdot \psi_n)$ increases to $|\phi|$. By the MCT, $\|\phi\|^* = \int |\phi| = |\int |\phi|\,| = \lim |\int \phi \cdot \psi_n| \le \epsilon$ and thus $\|f\|^* \le \|f - \phi\|^* + \|\phi\|^* \le 2\epsilon$. As ϵ was arbitrary, f is negligible. The answer to the second question is "yes." Hint: Use the Stone–Weierstraß theorem.

II.7.2 Let $\phi_n \in \mathcal{E}$ with $\sup_n \phi_n > 0$. The countable collection $[\phi_n > 1/k]$ of integrable sets covers the ambient space. Conversely, let $A_k, k = 1, 2, \ldots$ be integrable sets that cover the ambient space. By the very definition of Daniell's mean there exist $h_k = \sup_n \phi_k^{(n)} \in \mathcal{E}^{\uparrow}$ with $h_k \ge A_k$. Now rearrange the functions $0 \vee \phi_k^{(n)} \wedge 1$ in a sequence.

II.7.4 Let $\{I_k\}$ be a countable family of integrable sets whose union is the whole ambient space. If $\mathcal{M}$ were uncountable then one of the collections $\mathcal{M}_k = \{A \cap I_k : A \in \mathcal{M}\}$ would contain uncountably many non–negligible sets. For some $r \in \mathbb{N}$ we would have $\|A\|^* > 1/r$ for uncountably many $A \in \mathcal{M}_k$, which would contradict Lemma II.5.3 on p. 53, (iii).

II.7.9 $A = \bigcap_{N \in \mathbb{N}} \bigcup_{n \ge N} A_n$. If $\sum_n \int A_n < \infty$ then

$$\int \bigcup_{n \ge N} A_n \le \sum_{n \ge N} \int A_n \xrightarrow[N \to \infty]{} 0\,.$$

II.7.12 (i,ii): An open interval (a, b) is the pointwise dominated limit of step functions $\phi_n = (a, b - 1/n]$, so it is integrable, and its integral $\lambda((a, b))$ is the limit $b - a$ of the integrals $b - a - 1/n$ of the ϕ_n. If $U = \bigcup_i (a_i, b_i)$ is the canonical representation of the open set U, then $\lambda(U) = \sum (b_i - a_i)$ by Exercise II.7.10. (iii): Any compact set K is of the form $[-M, M] \setminus U$, U open, and so is integrable.
(iv): Let $h \in \mathcal{E}^{\uparrow}$ with $h \ge f$ and $\int^* h \le \int^* f + \epsilon/2$. There exists an increasing sequence (ϕ_n) of elementary functions whose pointwise supremum is h. The sets $[\phi_n > 1/2]$ are finite unions $\bigcup_{i=1}^{I(n)} (a_n^i, b_n^i]$ of elementary intervals. The set

$$U = \bigcup_{i,n=1}^{\infty} (a_n^i, b_n^i + \epsilon 2^{-n-i-1})$$

is open and contains A. It is easily seen that $\|U - A\|^* \le 2\epsilon$.
(v): There is an M such that $\lambda(A \cap [-M, M]) \ge \lambda(A) - \epsilon/2$. It suffices to find K compact inside $A \cap [-M, M]$ with $\lambda(A \cap [-M, M] \setminus K) \le \epsilon/2$. In other words, we may assume that A lies inside some interval $[-M, M]$. By (iii), there is an open set U containing $[-M, M] \setminus A$ and differing from this set in measure by less than ϵ. $K = [-M, M] \setminus U$ meets the description.

II.8.1 Let $f_n = (-1, 1) \setminus (\bigvee_{k=1}^n B_k \cap (-1, 1))$.

III.1.9 The algebra generated by $\mathcal{C}$ is dense in $\mathcal{L}^1$. The $\mathcal{C}$–uniformity agrees on $[-n, n]$ with the usual uniformity of $\mathbb{R}$. By the Localization Principle III.3.5, it suffices to check a function for measurability on each of these intervals.

III.2.7 Let the integable set A and $\epsilon > 0$ be given. The proof of Lemma III.2.1 applies and produces an integrable subset $A_1 \subset A$ with $\|A - A_1\|^* < \epsilon/2$ on which everyone of the functions f_n is $\mathcal{E}$–uniformly continuous. Note now that the functions $\psi_{m,n}(x) = d(f_m(x), f_n(x))$ are $\mathcal{E}$–uniformly continuous on A_1. By Theorem I.3.25, everyone of them is, on A_1, the sum of a constant and a function of $\overline{\mathcal{E}}$. From here on we copy the proof of Observation III.1.2. We conclude that $A_1 \cdot \psi_{m,n}$ is integrable. Then so is the set (see Proposition II.7.1)

$$B_p^r = A \cap \bigcup_{m,n \geq p} \left[\psi_{m,n} > \frac{1}{r} \right].$$

As p increases, B_p^r decreases, and the intersection $\bigcap_p B_p^r$ is contained in the negligible set of points where (f_n) does not converge. Continue as on page 71.

III.2.8 If f is $\mathcal{E}$–uniformly continuous on A_0 then $\phi \circ f$ is uniformly continuous there as well in case (a). In case (b) the image $f(A_0)$ is relatively compact, ϕ is uniformly continuous on it, and $\phi \circ f$ again is $\mathcal{E}$–uniformly continuous on A_0.

III.2.9 By Lemma III.2.1, f is measurable for the $\mathcal{C}$–uniformity on E. On the compact set $\overline{f(A_0)}$, this coincides with the d–uniformity.

III.3.9 Given an integrable set A and $\epsilon > 0$ we find a countable subcollection $\{C_1, C_2, \ldots\}$ of $\mathcal{C}$ whose union covers A up to a negligible set. The sets $C'_n = A \cap \bigcup_{k=1}^n C_i$ increase a.e. to A. By the MCT one of them, C'_n say, will have $\|A \backslash C'_n\|^* < \epsilon/2$. On it f is measurable, by Theorem III.3.5. There is therefore an integrable set $A_0 \subset C'_n$ with $\|C'_n - A_0\|^* < \epsilon/2$ on which f agrees with a function from $\overline{\mathcal{E}}$. The claim follows from the observation that $\|A \backslash A_0\|^* < \epsilon$.

III.3.10 Let $A \subset \mathbb{R}$ be integrable. Given $\epsilon > 0$, find a compact $K \subset A$ with $\|A - K\|^* < \epsilon/2$ (Proposition II.7.12) and cover K with finitely many neighborhoods on which f is measurable.

III.3.17 The proofs of the permanence properties of $\mathcal{F}_E^*$ and $\mathcal{L}_E^1$ are all the same as in the real or complex case; one merely replaces judiciously the absolute–value sign by $\| \ \|_E$.

III.3.18 Use Lemma II.2.14 on p. 39 to show that the E–valued step functions over $\mathcal{R}$ are exactly the sums of functions of the form $\xi \cdot A$, $\xi \in E, A \in \mathcal{R}$, and thus equal $\mathcal{E} \otimes E$. The argument will show that the integral of Equation (III.3.2) is given by Equation (III.3.3) and is thus well–defined. The majorization also is obvious from there. Let then (ϕ_n) be a sequence in $\mathcal{E} \otimes E$ that converges in the $\| \ \|^*$–norm to $f \in \mathcal{L}_E^1$. Since E is complete, the argument of page 58 applies and shows that the sequence $\left(\int \phi_n\right)$ has a unique limit $\int f \in E$. The linearity of the integral, and the fact that it is still majorized by $\| \ \|^*$ is shown literally as in Theorem II.6.1 on p. 58.

III.3.21 Let $\Phi = \{\phi \in \mathcal{E} : \ 0 \leq \phi \leq 1 \ , \ \|\phi \cdot A\|^\bullet = 0\}$. This is an increasingly directed set, and its supremum h is measurable. $\underline{A} \stackrel{\text{def}}{=} A \setminus [h > 0]$ meets the description.

III.3.22 Let $\mathcal{C}$ be a maximal collection of mutually disjoint non–negligible integrable sets C with the property that $C = \underline{C}$. Such exists by Zorn's lemma and meets the description.

III.4.6 This is immediate from Theorem III.7.6 and a bit of set algebra.

III.4.12 If not there would be an $\epsilon > 0$ and a subsequence with $\|f_{n_{k+1}} - f_{n_k}\|^* > \epsilon \quad \forall k$. There would be a further subsequence of it converging almost everywhere to f. This, however, contradicts Proposition II.5.20.

III.5.3 Let $\mathcal{E}_\wedge = \{f \in \mathcal{E}^\Sigma : f \wedge 1 \in \mathcal{E}^\Sigma\}$. This collection contains $\mathcal{E}$ and is sequentially closed. It therefore coincides with $\mathcal{E}^\Sigma$.

III.5.6 The argument has been given twice already: in Proposition II.7.5 on p. 63 and in Proposition III.4.4 on p. 82.

III.5.10 Let us show, for example, that it is closed under taking sums. Let $f, g \in \mathcal{M}(\mathcal{F})$. The set $[f + g > r]$ can be written as $\bigcup_{q \in \mathbb{Q}} [f > q] \cap [g > r - q]$, showing that it belongs to $\mathcal{M}(\mathcal{F})$. Multiplication is handled similarly, the other properties are even simpler to show.

III.5.15 This is true if ϕ is continuous, and the class of functions ϕ meeting this description is sequentially closed.

III.5.17 $\mathcal{M}$ clearly contains the algebra $\mathcal{A}$ generated by $\mathcal{I}$ and 1. It also contains the uniform closure $\overline{\mathcal{A}}$ of $\mathcal{A}$ (Exercise I.3.17). By Proposition I.3.10 on p. 11, $\overline{\mathcal{A}}$ is a vector lattice. The monotone class $\overline{\mathcal{A}}^m$ generated by $\overline{\mathcal{A}}$ belongs to $\mathcal{M}$ and is a vector lattice. To see this consider the collection

$$C^\wedge = \{f \in \overline{\mathcal{A}}^m : f \wedge \phi \in \overline{\mathcal{A}}^m \quad \forall \phi \in \overline{\mathcal{A}}\} .$$

This is clearly a monotone class containing $\overline{\mathcal{A}}$, so it contains $\overline{\mathcal{A}}^m$. Next consider the collection

$$\{g \in \overline{\mathcal{A}}^m : g \wedge f \in \overline{\mathcal{A}}^m \quad \forall f \in \overline{\mathcal{A}}^m\} .$$

This is again a monotone class containing $\overline{\mathcal{A}}$ so it contains $\overline{\mathcal{A}}^m$. The upshot: $\overline{\mathcal{A}}^m$ is closed under taking infima. Similarly, $\overline{\mathcal{A}}^m$ is closed under finite suprema, (linear combinations, products, etc). A monotone class closed under infima and suprema is evidently closed under pointwise converging sequences ($\lim f_n = \sup_N \inf_{n>N} f_n$). The upshot is that $\overline{\mathcal{A}}^m \subset \mathcal{M}$ is a sequentially closed lattice algebra containing $\mathcal{I}$. It contains therefore the σ–algebra generated by $\mathcal{I}$ and every bounded function measurable on it.

III.5.21 The definition of $\mathcal{F}/\mathcal{G}$–measurability says that this is true if f is (the indicator function of) a set in $\mathcal{G}$. The collection of f for which it is true is a sequentially closed lattice ring, so it contains all $\mathcal{G}$–measurable functions.

III.6.4 In general, no. This is hard.

III.7.1 Apply the proof of the σ–additivity of μ_F on page 143 to the function $F(x) = x$ and the sequence $(A \setminus A_n)$.

IV.1.2 The function $\phi(t) = (1-\lambda)+\lambda{\cdot}t - t^\lambda$ has derivative $\phi'(t) = \lambda(1-t^{\lambda-1}) \le 0$ and second derivative $\lambda(1-\lambda)t^{\lambda-2} \ge 0$. It is therefore convex and has its minimum at $t = 1$. It is, in particular, positive on $(0,\infty)$. Setting $t = a/b$ results in the desired inequality. If $b = 0$ there is nothing to prove.

IV.1.3 For $1 < p < \infty$ and $g = f{\cdot}|f|^{p-2}$, equality obtains in Hölder's inequality. If $p = 1$, equality obtains for $g = \operatorname{sgn} f$. If $p = \infty$ suppose $\|f\|_\infty > r$ and let A be a non–negligible subset of $[|f| > r]$ of finite outer measure; then set $g = A/\lambda^*(A)$. Note that g can be chosen measurable or positive if f is.

IV.1.6 It has to be shown that, for any two p–integrable functions f, g with $0 \le f \le g$ a.e., $\|f\|_p = \|g\|_p$ implies $f = g$ a.e. (III.6.6). If $p = 1$ this is clear, since $\|\ \|_1 = \|\ \|^*$, and $\|\ \|^*$ has this property. Consider then the case $1 < p < \infty$: $0 \le f \le g \implies |f|^p \le |g|^p$. Both $|f|^p$ and $|g|^p$ are 1–integrable functions (Theorem IV.2.5); so if $\|f\|_p = \|g\|_p$ then $f^p = g^p$ a.e. and $f = g$ a.e.

IV.2.12 It is the vector space generated by $\overline{\mathcal{E}}$ and the negligible functions. It consists exactly of the functions $\overline{\phi} + \nu$ with $\overline{\phi} \in \overline{\mathcal{E}}$ and $\|\nu\|^* = 0$.

IV.2.14 After suitable modification all but Exercises II.6.7, II.6.2, II.6.12, II.6.3, II.6.9, II.7.14, and II.7.16, which refer directly to the integral and its linearity.

IV.2.15 Use Exercise I.7.15.

IV.3.3 $v \mapsto v + w$ is an order isomorphism. $b \ge v$ iff $-v = b + (-b-v) \ge v + (-b-v) = -b$; thus $b \ge v \quad \forall v \in \mathcal{F}$ iff $-b \le v' \quad \forall v' \in -\mathcal{F}$.

IV.3.9 Let B be the non–integrable set of Section II.8, and let $\mathcal{B}$ be the collection $\{\{x\} : x \in B\}$ of functions, each of which vanishes in all but one point (Convention I.2.6 on p. 6). $\mathcal{B}$ has many upper bounds in $\mathcal{L}^p$, but not a smallest one.

IV.4.8 ϕ takes its maximum at some point $k \in K$. Set $\mu(\psi) = \psi(k)$ for $\psi \in C_{00}[K]$. (This measure μ is called ***Dirac measure*** at k.)

IV.4.10 Regard $\| \ \|^*$ as a mean on the lattice ring $\mathcal{E}' \stackrel{\text{def}}{=} \mathcal{L}^1_b[\mathcal{E}, \| \ \|^*]$, and let $\mathcal{B}$ denote the collection of all pairs $g, g' \in \mathcal{L}^1[\| \ \|^*]$ with $g \le g'$. For every $(g, g') \in \mathcal{B}$ let $m_{g,g'}$ be a positive linear functional on $\mathcal{L}^1$ with $\|m_{g,g'}\|_{\mathcal{L}^{1*}} \le 1$ and $\langle m_{g,g'}|g'-g\rangle = \|g'-g\|^*$. Then set

$$\|f\|^\sharp \stackrel{\text{def}}{=} \|f\|^* + \epsilon \cdot \sup\left\{ \langle m^*_{g,g'}|f\rangle : (g, g') \in \mathcal{B} \right\} , \qquad f : S \to \overline{\mathbb{R}} .$$

Here $m^*_{g,g'}$ is of course the Daniell mean of the σ–additive positive elementary integral $m_{g,g'} : \mathcal{E}' \to \mathbb{R}$. Clearly $\| \ \|^*$ and $\| \ \|^\sharp$ have the same negligible sets, and Inequality (IV.4.3) is satisfied for $f \in \mathcal{E}'$ and therefore also on the closure $\mathcal{L}^1[\| \ \|^*] = \mathcal{L}^1[\| \ \|^\sharp]$. Now $\| \ \|^\sharp$ is strictly increasing. For if $f \le g$ in $\mathcal{L}^1$ and $\|f\|^\sharp = \|g\|^\sharp$ then, since every ingredient in $\| \ \|^\sharp$ is increasing, we must have $\langle m^*_{f,g}|f\rangle = \langle m^*_{f,g}|g\rangle$ and thus

$$\|g - f\|^* = \langle m_{f,g}|(g-f)\rangle = \langle m^*_{f,g}|g\rangle - \langle m^*_{f,g}|f\rangle = 0 ,$$

which implies $\|g - f\|^\sharp = 0$.

IV.5.3 If $\langle \mathbf{g}^*|A\rangle \ge 0$ for all integrable sets A then $\langle \mathbf{g}^*|f\rangle \ge 0$ for all positive step functions over integrable sets. These are dense in $\mathcal{L}^p$. Given $f \in \mathcal{L}^p_+$, we can find a sequence (f_n) of positive integrable step functions converging in p–mean to f. By the continuity of $\mathbf{g}^*$ in p–mean, $\langle \mathbf{g}^*|f\rangle = \lim\langle \mathbf{g}^*|f_n\rangle \ge 0$.

IV.6.8 We may assume S is countable (Why?). If $e_1, \ldots, e_N$ have been constructed, look for an element $f \in S$ such that $f - \sum_{n=1}^N \langle e_n|f\rangle \cdot e_n$ is not zero. If there is, normalize it and call it e_{N+1}. For the second statement use for S the collection $\{e^{-mx^2}x^n : m, n \in \mathbb{N}\}$.

V.1.2 Instead of doing this directly with the Stone–Weierstraß theorem, one can simply apply the subsequent integration theory of the product, which applies to elementary integrals defined on rings of functions.

V.1.3 With ϕ as in (V.1.1),

$$\phi^+ = \sum_i \left(\phi^X_{i+}\phi^Y_{i+} + \phi^X_{i-}\phi^Y_{i-} \right) ,$$

$$\phi^- = \sum_i \left(\phi^X_{i+}\phi^Y_{i-} + \phi^X_{i-}\phi^Y_{i+} \right) ,$$

$$\psi = \sum_i |\phi^X_i| \cdot |\phi^Y_i| .$$

V.2.4 Use lower and upper $\mathcal{E}$–Baire envelopes $\underline{f}$ and $\overline{f}$ for f. $\underline{f}(x) - t$ and $\overline{f}(x) - t$ are $\mathcal{E} \otimes \mathcal{E}[\mathbb{R}]$–Baire and by Theorem V.2.2 $\overline{f} - \underline{f}$ is $\mu \otimes \lambda$–negligible.

V.4.4 Let $2 < p < \infty$. The conjugate exponent $p' = p/(p-1)$ lies in the interval $(1, 2)$, and by Theorem V.4.3

$$\int U(\phi) \cdot \psi = \int \phi \cdot U(\psi) \le \|\phi\|_p \cdot \|U(\psi)\|_{p'} \le \|\phi\|_p \cdot A_{p'}\|\psi\|_{p'} .$$

We take the supremum over ψ with $\|\psi\|_{p'} \le 1$ and arrive at

$$\|U(\phi)\|_p \le A_{p'} \cdot \|\phi\|_p \,.$$

The claim is satisfied with $A_p = A_{p'}$. Note that, by interpolating now between $3/2$ and 3, say, the family of constants A_p can be chosen so it has no pole at $p=2$.

V.5.2 Again this is best done in two steps.
1) $¦m¦$ *is subadditive:* Let $\psi_1, \psi_2 \in \mathcal{E}_+$ and assume $¦m¦(\psi_1+\psi_2) > r$. There exists an $\phi \in \mathcal{E}$ with $|\phi| \le \psi_1 + \psi_2$ and $|m(\phi)| > r$. Now

$$\psi_1' = |\phi| \wedge \psi_1 \le \psi_1 \quad \text{and} \quad \psi_2' = |\phi| - |\phi| \wedge \psi_1 \le \psi_2 \,.$$

Set
$$\begin{array}{rclrcl} \phi_{1+} &=& \psi_1' \wedge \phi_+ \,, & \phi_{2+} &=& \phi_+ - \psi_1' \wedge \phi_+ \,, \\ \phi_{1-} &=& \psi_1' - \psi_1' \wedge \phi_+ \,, & \phi_{2-} &=& \phi_- + \psi_1' \wedge \phi_+ - \psi_1' \,. \end{array}$$

These are positive elementary integrands. The columns of this matrix add up to ψ_1' and ψ_2', the rows to ϕ_+ and ϕ_-. We estimate

$$\begin{aligned} r < |m(\phi)| &= |m(\phi_{1+} - \phi_{1-}) + m(\phi_{2+} - \phi_{2-})| \\ &\le ¦m¦(\psi_1') + ¦m¦(\psi_2') \le ¦m¦(\psi_1) + ¦m¦(\psi_2) \,. \end{aligned}$$

Consequently $¦m¦(\psi_1+\psi_2) \le ¦m¦(\psi_1) + ¦m¦(\psi_2)$: the subadditivity is established.
2) $¦m¦$ *is superadditive:* Let $\psi_1, \psi_2 \in \mathcal{E}_+$ with $¦m¦(\psi_1) + ¦m¦(\psi_2) > r$. Then there are $\phi_1, \phi_2 \in \mathcal{E}$ with $|\phi_i| \le \psi_i$ and $|m(\phi_1)| + |m(\phi_2)| > r$. Replacing if necessary ϕ_i by $-\phi_i$, we can see to it that $m(\phi_i)$ is positive and consequently $m(\phi_1+\phi_2) > r$. Now $\phi_1+\phi_2$ is an elementary function with $|\phi_1+\phi_2| \le \psi_1+\psi_2$, so we conclude that $¦m¦(\psi_1+\psi_2) > r$: the superadditivity is established.

V.5.6 The additivity is best shown in two steps.
1) $¦\mu¦$ *is subadditive:* Let $A_1, A_2 \in \mathcal{A}$ with $A_1 + A_2 \in \mathcal{A}$ (i.e., disjoint), and assume $¦\mu¦(A_1+A_2) > r$. There exist $B^1, B^2 \in \mathcal{A}$ with sum $A_1 + A_2$ and $\mu(B^1) - \mu(B^2) > r$. Set

$$\begin{array}{rclrcl} B_1^1 &=& A_1 \wedge B^1 \,, & B_2^1 &=& B^1 - A_1 \wedge B^1 \,, \\ B_2^1 &=& A_1 - A_1 \wedge B^1 \,, & B_2^2 &=& A_2 + A_1 \wedge B^1 - B^1 \,. \end{array}$$

These four sets belong to $\mathcal{A}$. This is obvious for all but B_2^2; and for this it follows from $A_2 + A_1 \wedge B^1 - B^1 = (A_1 + A_2 - B^1) \wedge A_2 = B^2 \wedge A_2$. The columns of this matrix add up to A_1 and A_2, the rows to B^1 and B^2. Evidently

$$\begin{aligned} ¦\mu¦(A_1) + ¦\mu¦(A_2) &\ge \Big(\mu(B_1^1) - \mu(B_1^2)\Big) + \Big(\mu(B_2^1) - \mu(B_2^2)\Big) \\ &= \mu(B_1^1 + B_2^1) - \mu(B_1^2 + B_2^2) = \mu(B^1) - \mu(B^2) > r \ : \end{aligned}$$

the subadditivity is established.
2) $¦\mu¦$ *is superadditive:* Let $A_1, A_2 \in \mathcal{A}$ with $¦\mu¦(A_1) + ¦\mu¦(A_2) > r$. Then there are $B_i^1, B_i^2 \in \mathcal{A}$ with $A_i = B_i^1 + B_i^2$ and

$$\Big(\mu(B_1^1) - \mu(B_1^2)\Big) + \Big(\mu(B_2^1) - \mu(B_2^2)\Big) > r \,.$$

This reads $\mu(B^1) - \mu(B^2) > r$, where $B^1 = B_1^1 + B_2^1$ and $B^2 = B_1^2 + B_2^2$ are sets of $\mathcal{A}$ with sum $A_1 + A_2$. We conclude that $¦\mu¦(A_1 + A_2) > r$: The superadditivity is established as well, and with it the additivity.

(iii): The condition that μ be σ–additive is clearly necessary. Namely, if $\mathcal{A} \ni A_n \downarrow 0$ and $¦\mu¦$, $\int d\mu$, or $\int d¦\mu¦$ are σ–additive then $|\mu(A_n)| = |\int A_n \, d\mu| \le ¦\mu¦(A_n) = \int A_n \, d¦\mu¦ \to 0$.
Suppose it is satisfied, and let (A_n) be an increasing sequence in $\mathcal{A}$ with union $A \in \mathcal{A}$. If $¦\mu¦(A) > r$ then there are disjoint sets $A^1, A^2 \in \mathcal{A}$ with union A and $\mu(A^1) - \mu(A^2) > r$. The sequences $(A^1 \cap A_n)$ and $(A^2 \cap A_n)$ increase to A^1 and A^2, respectively. By the σ–continuity of μ, $\mu(A^1 \cap A_n) - \mu(A^2 \cap A_n) > r$ for sufficiently large indices n, which implies $¦\mu¦(A_n) > r$. Thus $¦\mu¦$ is σ–additive (See Exercise II.2.9 on p. 38).
Next let (ϕ_n) be a sequence in $\mathcal{E}[\mathcal{A}]$ that decreases pointwise to 0. Let $\epsilon > 0$. The sets $[\phi_n > \epsilon] \in \mathcal{A}$ decrease pointwise to the void set. There is an $N \in \mathbb{N}$ such that $¦\mu¦([\phi_n > \epsilon]) < \epsilon$ for $n \ge N$. For such n

$$
\begin{aligned}
\int \phi_n \, d¦\mu¦ &= \int \phi_n \cdot [\phi_n > \epsilon] \, d¦\mu¦ + \int \phi_n \cdot [\phi_n \le \epsilon] \, d¦\mu¦ \\
&\le \int \phi_1 \cdot [\phi_n > \epsilon] \, d¦\mu¦ + \int \epsilon \cdot [\phi_n > 0] \, d¦\mu¦ \\
&\le \|\phi_1\|_u \, ¦\mu¦([\phi_n > \epsilon]) + \epsilon \cdot ¦\mu¦([\phi_1 > 0]) \\
&\le \epsilon \left(\|\phi_1\|_u + ¦\mu¦([\phi_1 > 0]) \right) .
\end{aligned}
$$

This shows that $\int d¦\mu¦$ is σ–additive, and then clearly so is $\int d\mu$.

V.6.15 $¦m¦ = \sum_{n=1}^{\infty} |r_n| \cdot \delta_{x_n}$.

V.9.4 (i): The linearity and positivity are evident. The contractivity follows from (iii) and the observation that $x \mapsto |x|^p$ is convex when $1 \le p < \infty$.
(ii): There is a countable collection of linear functions $\ell_n(x) = \alpha_n + \beta_n x$ such that $\Gamma(x) = \sup_n \ell_n(x)$ at every point $x \in \mathbb{R}$. Linearity and positivity give

$$
\ell_n(\mathbb{E}^\mu[f|\Phi]) = \mathbb{E}^\mu[\ell_n(f)|\Phi] \le \mathbb{E}^\mu[\Gamma(f)|\Phi] \qquad \text{a.s.} \quad \forall n \in \mathbb{N} .
$$

Upon taking the supremum over n, Jensen's Inequality (V.9.1) follows.

V.10.6 Apply Lemma V.10.5 to $E \cap (n-1, n)$ and add.

V.10.7 Apply it to $[-n, n] \cdot g$ and take the limit.

References

1. Bichteler, K.
Integration theory, Lecture Notes in Mathematics **315**, Springer, Berlin, Heidelberg, New York, 1973.

2. Bourbaki, N.
Intégration, Hermann, Paris, 1965.

3. Breiman, L.
Probability, Addison-Wesley, Reading Mass., 1968.

4. Carleson, L.
"On the convergence and growth of partial sums of Fourier series," Acta Math 116 (1966), 135–157.

5. Du Bois–Reymond, P.
"Untersuchungen über die Convergenz und Divergenz der Fourierschen Darstellungsformeln," Abhandlungen der Akademie München XII (1876), 1–103.

6. Halmos. P.
Measure Theory, Van Nostrand, New York, 1950.

7. Kolmogorov, A. N.
"Une séries de Fourier-Lebesgue divergente partout," C.R.Acad Sci. Paris, 183 (1926), 1327-28.

8. Riesz, M.
"Sur les ensembles compact de fonctions sommable," Acta Sci. Math. Szeged 6 (1933) 136-42.

9. Royden, H. L.
Real Analysis, 2nd edition, Macmillan, NewYork 1968.

10. Thomas, E. G. F.
"L'integration par rapport a une mesure de Radon vectorielle," Ann. Inst. Fourier 20 (1970), 55–191.

K. Bichteler, *Integration – A Functional Approach*, Modern Birkhäuser Classics,
DOI 10.1007/978-3-0348-0055-6, © Birkhäuser Verlag 1998

Index of Notations

Symbols

Index

Boldface pagenumbers refer to definitions.

I

J

L

Breinigsville, PA USA
26 December 2010
252124BV00003B/36/P

9 783034 800549